Demystifying Numerical Models

Demystifying Numerical Models
Step-by-Step Modeling of Engineering Systems

John P. T. Mo
Manufacturing Engineering, RMIT University,
Melbourne, VIC, Australia

Sherman C. P. Cheung
Mechanical Engineering, RMIT University,
Melbourne, VIC, Australia

Raj Das
Aerospace and Aviation,
RMIT University, Melbourne, VIC, Australia

Butterworth-Heinemann
An imprint of Elsevier

Butterworth-Heinemann is an imprint of Elsevier
The Boulevard, Langford Lane, Kidlington, Oxford OX5 1GB, United Kingdom
50 Hampshire Street, 5th Floor, Cambridge, MA 02139, United States

Notices
Knowledge and best practice in this field are constantly changing. As new research and experience
broaden our understanding, changes in research methods, professional practices, or medical treatment
may become necessary.

Practitioners and researchers must always rely on their own experience and knowledge in evaluating and
using any information, methods, compounds, or experiments described herein. In using such information
or methods they should be mindful of their own safety and the safety of others, including parties for
whom they have a professional responsibility.

To the fullest extent of the law, neither the Publisher nor the authors, contributors, or editors, assume any
liability for any injury and/or damage to persons or property as a matter of products liability, negligence
or otherwise, or from any use or operation of any methods, products, instructions, or ideas contained in
the material herein.

British Library Cataloguing-in-Publication Data
A catalogue record for this book is available from the British Library

Library of Congress Cataloging-in-Publication Data
A catalog record for this book is available from the Library of Congress

ISBN: 978-0-08-100975-8

For Information on all Butterworth-Heinemann publications
visit our website at https://www.elsevier.com/books-and-journals

Working together
to grow libraries in
developing countries

www.elsevier.com • www.bookaid.org

Publisher: Matthew Deans
Acquisition Editor: Brian Guerin
Editorial Project Manager: Leticia Lima
Production Project Manager: Anitha Sivaraj
Cover Designer: Victoria Pearson

Typeset by MPS Limited, Chennai, India

Contents

Preface

Engineering is the application of scientific knowledge to create systems that serve people and the community. In the engineering design process, engineers use data to characterize, refine, test, validate, verify, and plan system functions and behaviors. However, most engineering design literatures focus on external esthetic design capabilities and the use of graphical systems to create and process the product. Texts that focus on functional and behavioral designs are analytic making it difficult for the students to comprehend without going through the tedious mathematical derivations. The problem is that students tend to memorize the mathematical expression without an in-depth understanding of the fundamental principles in engineering.

For some physical analysis (e.g. structural and fluid flow behavior), specific computational packages specially designed for the type of assets have been developed. These software packages are designed for detailed analysis for large-scale sophisticated systems with specific applications. Nonetheless, in practical engineering process, engineers are constantly facing the challenges to assess system behavior characterized by differential/integral mathematical expression where analytical solution is not accessible; such as logistics and services, or even human behavior operating an engineering asset. Computational tools for obtaining numerical solutions of these problems are simply not available. The engineers have to rely on their own knowledge and methodology to design their system.

This book fills the gap by explaining the analytic concepts of engineering components and systems in simplified terms and progressing to focus on solving the engineering characteristics and behaviors using numerical methods. The computational aspects of engineering analysis can then be applied to develop the functions and characteristics of the engineering systems to a level that is adequate for implementation.

This book aims to target audience both for undergraduates studying engineering and practical engineers in industry. For undergraduate audience, basically, all engineering disciplines will need this knowledge for solving complex problems that are analytically impossible to derive a solution or for projects that do not show strong correlation in operating parameters. Worked examples are used extensively to illustrate different methods and approaches.

Students studying systems engineering, mechanical engineering, manufacturing engineering, mechatronics, industrial engineering, infrastructure planning, process engineering, electrical and power engineering will find the examples particularly relevant.

For practical engineers who understand the principles and have a broad range of experience in engineering, they do not want to be constrained by the mathematical formulation of their problems. Instead, they can adapt the numerical models presented in this book to develop a holistic view of the engineering system's performance.

This book is recommended to classes:

- Engineering Numerical Methods
- Fluid Mechanics and Thermodynamics
- Vibrations and Controls
- Solid Mechanics
- Risk assessment and analysis
- Systems Engineering Principles
- System Reliability and Services

This book draws on examples in many daily life engineering issues. For example, water supply and sewage treatment touch on everybody's life. These systems work on the principles of fluid flows. They are very complex and require a lot of engineering analysis to ensure a balance. The waste water treatment system is even more complicated. This type of systems is traditionally analyzed and designed separately in different components. Integration of these components to a complete system depends on the knowledge that engineers bring along after years of working experience in the field. This book is written in such a way that practical experience is represented in numerical forms with interactive spreadsheets making the knowledge more readily learnt by the readers.

This book naturally competes with traditional systems engineering books. A good example is the book by Benjamin Blanchard and Walter Fabrychy, "Systems Engineering and Analysis." This book focuses on the modeling of systems and has a lot of conceptual development of models. However, when it comes to actual design of the "system," the technical content is often insufficient to support a detail engineering analysis, due to the difficulty of the mathematical formulation and analytical solutions. This book supplements the gap by extending the high-level modeling practice to integrating the "system" numerically into a testable model. Using established numerical means, solutions to the problem can be developed with the help of modern computational platforms.

One of the main objectives of this book is to provide hands-on numerical examples and graphical visualization of complex mathematics through simple spreadsheets. Spreadsheets will be developed for each chapter to demonstrate the applications and numerical examples for a given practical problem. Readers could then interact with the downloadable spreadsheets, understand the relevant numerical techniques, and grasp an in-depth knowledge of the system behavior instead of solving the complex mathematics analytically.

John Mo, Sherman Cheung and Raj Das
December 2017

Chapter 1

Introduction to Engineering Systems

1.1 SYSTEMS ENGINEERING PRINCIPLES

A system is a unified set of components with different functionality working together toward a common goal. Not surprisingly, bringing many components together and aligning the varying functions toward the goal is already a challenge. Engineers have to overcome many challenges in the design and analysis of engineering systems. The following sections discuss critical grand challenges that an engineering system should be analyzed against.

1.1.1 Integrity

Engineers have the responsibility to create, design, manufacture, manage, and dispose of systems that operate safely, reliably, and with minimal negative impact to the society. Human lives can depend upon the quality of engineering project outcomes, and significant economic and environmental consequences can result from underperforming engineering facets. The concept of integrity in systems is to maintain total knowledge of the principles, characteristics, constraints, and processes that exist around the engineering system, so that any foreseeable problems can be prevented and any damages can be minimized, even in extreme circumstances. Numerical analysis helps to analyze integrity of engineering systems irrespective of whether they are linear, nonlinear, discretional, random, or any difficult to express characteristics.

1.1.2 Stability

The concept of stability originates from mechanics and structures. When mechanical structures are stable, all forces in the structures are in equilibrium such that loads are distributed to the structural members that can bear the load for a long time. Similarly, in other engineering branches such as electrical systems, a stable electrical circuit is one that has equilibrium of voltages and currents being distributed appropriately. An extension of this concept to systems is the maintenance of equilibrium condition, so that the

Demystifying Numerical Models. DOI: https://doi.org/10.1016/B978-0-08-100975-8.00001-1

system can operate and perform at the right level for a long period. Numerical methods can search through a broad range of variables in different scenarios to ascertain stability of the system during operations and extreme circumstances.

1.1.3 Compatibility

A compatible system is one that can exist and operate in a harmonious, agreeable, or congenial manner with other systems. Compatibility can occur in many ways. For example, for medical systems, a cochlear implant should be able to exist in the patient's body in a chemically and biochemically stable state, and stays harmoniously with other parts of the body. Modeling analysis can highlight incompatible interfaces between components and is the first step in resolving this problem.

1.1.4 Safety

Fatality and many types of injuries people operating engineering systems are irreversible. It is important to make sure that systems will operate such that personnel using or staying nearby the system are not exposed to any danger. Large-scale engineering systems operating in an extreme environment is no doubt much more dangerous. Any minor error can escalate to disaster easily if not carefully managed. When designing such a system, many safety measures must be installed, and processes are defined and rehearsed to ensure that these safety measures are followed. Each of these measures should be analyzed to ensure even the extreme situation will not induce significant safety issues.

1.1.5 Sustainability

Large complex engineering systems require huge investments from the stakeholders. It is clear that such a system is not supposed to serve its purpose for a short time only. This kind of systems are expected to be in-service for 30 years, and often longer. If 30 years is the average number of years for a generation, it is common that such complex systems are still in operation after a couple of generations of working personnel. During this time, many changes can take place. For example, technology may change so that the system on board becomes incompatible with ground systems, or some components are worn out after many cycles of operations (the so-called ageing effect). These changes can happen much sooner than the expected service life. Sustainability must be designed into the system, so that appropriate maintenance and upgrade services can be done in timely fashion. Sustainability design can be analyzed based on operating parameters and system model that is usually not readily represented by analytical models. The use of numerical analysis is an obvious choice to project system performance over this long period.

1.2 NATURE OF ENGINEERING SYSTEMS

Developers of complex engineering products such as an aircraft or a ship are facing new challenges in meeting business goals and competition globally. They need to remain competitive by developing innovative products and processes which are specific to individual customer's requirements, completely packaged, and made available globally to make best use of resources within defined constraints. New operational requirements demand not only a functional system, but also a reliable and precise product.

The complexity of these engineering products also means the need for full understanding and predictability of the system. However, many systems are working on the principles of nonlinear, and sometimes discontinuous or piecewise models. Analysis of these systems by solving equations becomes very difficult because more and more independent variables are incorporated into the system.

Professional engineers and operation managers working in the new engineering environment tend to use modern computational tools to analyze these systems. Typical process is to adopt whatever available components that produce the required performance outcomes. However, the overall performance of systems would not be predictable. Numerical methods are techniques that analysis the system with numbers. To examine the core knowledge base, we will use a number of examples to illustrate the key knowledge elements.

Chapter 2

Basic Numerical Techniques

2.1 INTRODUCTION

The basic concept of solving engineering system numerically is to express the system with a set of related numbers that will change as the independent variable of the system changes. The relationship of the set of numbers can be expressed in many forms. The most common form is to derive the analytical formula of the system using mathematical concepts. The analytical equations can then be expressed in a standard form that has known procedure developed by previous researchers to compute the behavior of the function.

There are some other types of engineering systems that are not possible to be expressed as analytical equations. An example is the procedure to find the optimal solution of layouts that can minimize transport times during operation. Another example is the development of reliability from event trees that only have logical relationships rather than through a continuous function. The numerical methods to solve these engineering problems are described in the specific chapters.

Whether it is a standard numerical procedure for analytical functions or a graphical relationship expressed in some kind of logic process, a common starting point of numerical techniques is to work with a set of initial values. This is in contrast to generalized engineering problems that express the outcomes as a function of the input variables. This does not mean that numerical techniques are one-off solutions. In fact, it is the procedure of working through the numerical solution that represents the generalized formulation of the engineering problems.

This chapter will introduce some of the main numerical methods that are useful for the engineering systems solutions later in this book.

2.2 ROOTS OF EQUATIONS

The solution to many engineering problems requires solving a set of equations which can be linear or nonlinear. The meaning of solving the equations is to find a set of values of the independent variables that make both sides of the equations equal. Several numerical methods are available to solve equations, all of them deal with single variable. Solutions to equations with multiple variables are limited to linear systems. Linear equations are solved by

Demystifying Numerical Models. DOI: https://doi.org/10.1016/B978-0-08-100975-8.00002-3

forming the matrix and finding the eigenvalues of the determinant matrix. Nonlinear equations are solved by a variety of methods depending on the nature of the equations.

2.2.1   Direct Search Method

The principle of direct search is based on traditional algebraic methods. Let's consider a nonlinear equation expressed in the form:

$$f(x) = 0 \tag{2.1}$$

For simple quadratic functions of the form in Eq. (2.2), the solution can be obtained readily by completing squares for both sides of the equation.

$$f(x) = ax^2 + bx + c = 0 \tag{2.2}$$

The roots are:

$$x_1 = \frac{-b + \sqrt{b^2 - 4ac}}{2a} \quad \text{and} \quad x_2 = \frac{-b - \sqrt{b^2 - 4ac}}{2a} \tag{2.3}$$

The roots of third-order polynomial can also be found analytically but there is no general solution for higher order polynomials. It should be noted that the number of roots of a polynomial is the order of the polynomial. However, for other nonlinear functions, the number of roots within a defined interval varies.

The simplest method to find the roots of a function is the direct search method. Graphically, it means plotting the function over the defined interval. If the function crosses the x-axis (i.e., $y = 0$), the root is the x-intercept as shown in Fig. 2.1.

The root can then be found by expanding the details of x around the x-intercept point. The direct search method will find the roots of any function so long as it is not case (c) in Fig. 2.1.

The following polynomial is used to illustrate the computational process of the direct search method for the interval $x \in [-1, 1]$. The results are presented in Table 2.1.

$$f(x) = x^3 - x^2 - 10x + 2 = 0 \tag{2.4}$$

It should be noted that the initial search interval is set up as 0.1 from -1.0 but as it passes 0.1 to 0.2, $f(x)$ changes from 0.991 to -0.032. The root (zero point) is certainly in between 0.19 and 0.2. The search interval is then reduced to 0.01 from 0.1. Again, as it passes 0.19 to 0.2, $f(x)$ changes from 0.070759 to -0.032. The search interval is further reduced to 0.001. Table 2.1 presents that the smallest search interval used is 0.00001. The two shaded cells are the closest to the root of $f(x)$ at the 0.00001 accuracy.

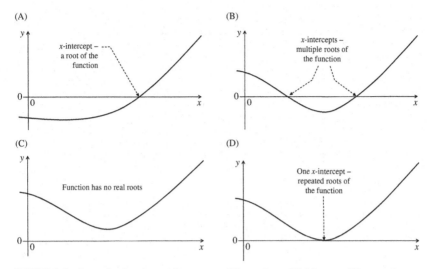

(A) x-intercept – a root of the function

(B) x-intercepts – multiple roots of the function

(C) Function has no real roots

(D) One x-intercept – repeated roots of the function

FIGURE 2.1 Roots for functions: (A) one root, (B) two (or multiple) roots, (C) no real roots, and (D) one repeated root.

To make it even more precise without going to another level of search, the root can be found by interpolating a point between 0.19688 and 0.19689 by:

$$f(x) = 0.19688 + 0.00001 \times \frac{0.000033}{0.0000697} = 0.196883 \qquad (2.5)$$

However, it is obvious that the method is cumbersome. The detail expansion aiming for high degree of precision requires very small subinterval sizes. More efficient method is preferred.

2.2.2 Bisection Method

The bisection method is modified from the direct search method such that the systematic procedure aims to eliminate some unnecessary expansion around the x-intercept. The method can be described in the following steps.

Step 1: Denote the desired interval in which a root is expected as $x \in [x_s, x_e]$, where $x_s < x_e$. Compute the midpoint of this interval:

$$x_m = \frac{x_s + x_e}{2} \qquad (2.6)$$

Step 2: Compute the value of the function at the start, end, and midpoint, i.e., $f(x_s)$, $f(x_e)$, and $f(x_m)$.

Step 3: Determine which case the zero point is going to be located (see Fig. 2.2).

Case 1: zero point lies between x_m and x_e, i.e.,

$\{f(x_s) > 0\}$ and $\{f(x_m) > 0\}$, or,

TABLE 2.1 Direct Search for f(x) in Interval [−1, 1]

x	f(x)	x	f(x)	x	f(x)	x	f(x)
−1.0	10.0	0.16	0.378496	0.1968	0.000892	0.19697	−0.00086
−0.9	9.461	0.17	0.276013	0.19681	0.000789	0.19698	−0.00096
−0.8	8.848	0.18	0.173432	0.19682	0.000686	0.19699	−0.00106
−0.7	8.167	0.19	0.070759	0.19683	0.000584	0.197	−0.00116
−0.6	7.424	0.191	0.060487	0.19684	0.000481	0.198	−0.01144
−0.5	6.625	0.192	0.050214	0.19685	0.000378	0.199	−0.02172
−0.4	5.776	0.193	0.03994	0.19686	0.000275	0.2	−0.032
−0.3	4.883	0.194	0.029665	0.19687	0.000172	0.3	−1.063
−0.2	3.952	0.195	0.01939	**0.19688**	**6.97E-05**	0.4	−2.096
−0.1	2.989	0.196	0.009114	**0.19689**	**−3.3E-05**	0.5	−3.125
0.0	2.000	0.1961	0.008086	0.19690	−0.00014	0.6	−4.144
0.1	0.991	0.1962	0.007058	0.19691	−0.00024	0.7	−5.147
0.11	0.889231	0.1963	0.00603	0.19692	−0.00034	0.8	−6.128
0.12	0.787328	0.1964	0.005003	0.19693	−0.00044	0.9	−7.081
0.13	0.685297	0.1965	0.003975	0.19694	−0.00055	1.0	−8.000
0.14	0.583144	0.1966	0.002947	0.19695	−0.00065		
0.15	0.480875	0.1967	0.00192	0.19696	−0.00075		

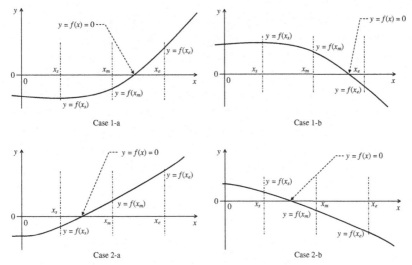

FIGURE 2.2   Case 1: root lies between x_m and x_e. Case 2: root lies between x_s and x_m.

$\{f(x_s) < 0\}$ and $\{f(x_m) < 0\}$
Case 2: zero point lies between x_s and x_m, i.e., not Case 1.
Step 4: Check the tolerance.

$$\text{For absolute value :} \varepsilon_a \geq \left| x_{m,i+1} - x_{m,i} \right| \tag{2.7}$$

$$\text{For relative value :} \varepsilon_r \geq \left| \frac{x_{m,i+1} - x_{m,i}}{x_{m,i+1}} \right| \tag{2.8}$$

Step 5: If the tolerance has been met, use x_m as the final estimate of the root. If the tolerance has not reached, continue:
For Case 1, set $x_s = x_m$. For Case 2, set $x_e = x_m$. From here, go back to Step 1 until the tolerance is satisfied.

The following polynomial is used to illustrate the computational process of the bisection method for the interval $x \in [-1, 1]$. The tolerance is set at $\varepsilon_a < 0.001$. The results are presented in Table 2.2.

$$f(x) = x^3 - x^2 - 10x + 2 = 0 \tag{2.9}$$

The true root cannot be solved analytically but a value to any accuracy can be found by direct search method, which obviously takes a lot of time. As shown by the direct search method, the true root is 0.196883. The iteration of bisection method toward true root is graphically shown in Fig. 2.3.

It should be noted that since $f(x)$ is a third-order polynomial, once one root is identified, the other two roots can be identified readily. Since the true

TABLE 2.2 Numerical Solution with Bisection Method at Interval [−1,1]

Iteration	x_s	x_e	x_m	$f(x_s)$	$f(x_e)$	$f(x_m)$	Case	ε_d
1	−1.0000	1.0000	0.0000	10.0000	−8.0000	2.0000	1	NA
2	0.0000	1.0000	0.5000	2.0000	−8.0000	−3.1250	2	0.5000
3	0.0000	0.5000	0.2500	2.0000	−3.1250	−0.5469	2	0.2500
4	0.0000	0.2500	0.1250	2.0000	−0.5469	0.7363	1	0.1250
5	0.1250	0.2500	0.1875	0.7363	−0.5469	0.0964	1	0.0625
6	0.1875	0.2500	0.2188	0.0964	−0.5469	−0.2249	2	0.0313
7	0.1875	0.2188	0.2031	0.0964	−0.2249	−0.0641	2	0.0156
8	0.1875	0.2031	0.1953	0.0964	−0.0641	0.0162	1	0.0078
9	0.1953	0.2031	0.1992	0.0162	−0.0641	−0.0240	2	0.0039
10	0.1953	0.1992	0.1973	0.0162	−0.0240	−0.0039	2	0.0020
11	0.1953	0.1973	0.1963	0.0162	−0.0039	0.0061	1	0.0010

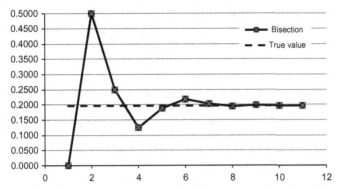

FIGURE 2.3   Comparison of bisection iteration values with true root value.

value of the function is not normally known, to derive the remaining roots, the last x_m (i.e., 0.1963) is used. Eq. (2.9) can be represented as:

$$f(x) = x^3 - x^2 - 10x + 2 = (x - 0.1963)(x^2 - 0.8037x + 10.1578) = 0 \quad (2.10)$$

The two other roots are found by Eq. (2.3):

$$x = 3.6142 \quad \text{or} \quad x = -2.8105$$

Note that each of these roots is subject to accuracy of the bisection outcome.

2.2.3   Newton–Raphson Method

Although the bisection method will always converge if the correct search interval is specified, the rate of converging to the root is slow. The Newton–Raphson method uses the slope of the function (i.e., the rate of change at the point which is in fact the first differential term in the Taylor series expansion) to fast track the search.

$$f(x_1) = f(x_0) + \left[\frac{df}{dx}(x_0) \right] \Delta x \quad (2.12)$$

where $f(x_1)$ is the value of the function at the incremented $x_1 = x_0 + \Delta x$ and $f(x_0)$ is the value at x_0. Not any incremental value Δx can be selected. The new incremented x_1 is expected to be close to the root, this means:

$$f(x_1) = f(x_0) + \left[\frac{df}{dx}(x_0) \right] (x_1 - x_0) = 0 \quad (2.13)$$

Re-arranging,

$$x_1 = x_0 - \frac{f(x_0)}{f'(x_0)} \quad (2.14)$$

where $f'(x_0) = (df/dx)(x_0)$. The new x_1 value is fully defined by x_0.

In general, at xi, the next interpolated $xi + 1$ is defined as:

$$x_{i+1} = x_i - \frac{f(x_i)}{f'(x_i)} \qquad (2.15)$$

Graphically, this is represented as shown in Fig. 2.4.

The same polynomial in Eq. (2.9) is used to illustrate the computational process of the Newton–Raphson method from the starting point of $x = -1$. The tolerance is set at $\varepsilon_a < 0.001$. First, the derivative function is derived:

$$f'(x) = 3x^2 - 2x - 10 \qquad (2.16)$$

The results are presented in Table 2.3.

In iteration $i = 1$, $x_i = -1$ as given and substitute into Eq. (2.9):

$$f(x_1) = x_1^3 - x_1^2 - 10x_1 + 2 = 10 \qquad (2.17)$$

Apply Eq. (2.16) for the derivative,

$$f'(x_1) = 3x_1^2 - 2x_1 - 10 = -5 \qquad (2.18)$$

Hence, from Eq. (2.15),

$$x_2 = x_1 - \frac{f(x_1)}{f'(x_1)} = -1 - \frac{10}{-5} = 1 \qquad (2.19)$$

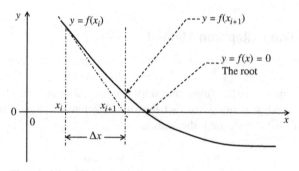

FIGURE 2.4 The principle of Newton–Raphson method.

TABLE 2.3 Numerical Solution With Newton–Raphson Method

Iteration i	x_i	$f(x_i)$	$f(x_{i+1})$	x_{i+1}	$f(x_{i+1})$
1	−1.0000	10.0000	−5.0000	1.0000	−8.0000
2	1.0000	−8.0000	−9.0000	0.1111	0.8779
3	0.1111	0.8779	−10.1852	0.1974	−0.0043
4	0.1973	−0.0043	−10.2778	0.1969	0.0000

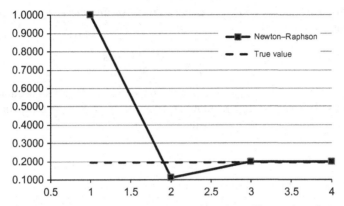

FIGURE 2.5   Comparison of Newton−Raphson iteration values with true root value.

The value of function at $x = x_2$ is then computed as,

$$f(x_2) = x_2^3 - x_2^2 - 10x_2 + 2 = 8 \tag{2.20}$$

Iteration 2 is repeated with the new $x_2 = 1$. The iteration process is much faster as shown in Fig. 2.5.

2.3   DIFFERENTIAL EQUATIONS

Several methods are available to solve differential equations. Whatever the method is, the differential equations should contain derivatives only, i.e., no integral terms. All methods work on the principle of adding incremental values to the function by evaluating the gradient at the start of each integration period. The resulting accuracy depends on how the slope at the incremental point is estimated.

2.3.1   Euler's Method

Euler's method uses the derivative to compute the value of the function $g(x)$ at next integration interval $x = x_0 + h$ as:

$$g(x) = g(x_0) + h\frac{dg}{dx}\bigg|_{x_0} \tag{2.21}$$

If the function $g(x)$ is more complex than the first-order derivative, i.e., the second and higher order derivatives are nonzero, there will be errors due to the higher order derivative terms. The errors can be estimated by expanding the function $g(x)$ by Taylor's series, so that the error term can be represented by:

$$e(x) = \frac{(x-x_0)^2}{2!}\frac{d^2g}{dx^2} + \frac{(x-x_0)^3}{3!}\frac{d^3g}{dx^3} + \cdots \tag{2.22}$$

It is necessary to note that the integration step size can affect accuracy of the results. This can be illustrated by an example. Consider the following differential equation with an initial value $g(0) = 1$:

$$\frac{dg}{dx} = 3x^2 \tag{2.23}$$

The analytical solution is obviously:

$$g(x) = x^3 + 1 \tag{2.24}$$

For an integration interval of 0.5, the Euler's method gives results in Table 2.4.

TABLE 2.4 Numerical Solution with Euler's Method at Interval 0.5

x_0	$\frac{dg}{dx}\big|_{x_0}$	$g(x_0)$	$g(x)$	Analytical Solution $g(x) = x^3 + 1$	Error	Error %
0.0	0.00	1.000	1.000	1.000	0	0
0.5	0.75	1.000	1.375	1.125	0.250	22.22
1.0	3.00	1.375	2.875	2.000	0.875	43.75
1.5	6.75	2.875	6.250	4.375	1.875	42.86
2.0	12.00	6.250	12.250	9.000	3.250	36.11
2.5	18.75	12.250	21.625	16.625	5.000	30.08
3.0	27.00	21.625	35.125	28.000	7.125	25.45
3.5	36.75	35.125	53.500	43.875	9.625	21.94
4.0	48.00	53.500	77.500	65.000	12.500	19.23
4.5	60.75	77.500	107.875	92.125	15.750	17.10
5.0	75.00	107.875	145.375	126.000	19.375	15.38
5.5	90.75	145.375	190.750	167.375	23.375	13.97
6.0	108.00	190.750	244.750	217.000	27.750	12.79
6.5	126.75	244.750	308.125	275.625	32.500	11.79
7.0	147.00	308.125	381.625	344.000	37.625	10.94
7.5	168.75	381.625	466.000	422.875	43.125	10.20
8.0	192.00	466.000	562.000	513.000	49.000	9.55

(Continued)

TABLE 2.4 (Continued)

x_0	$\frac{dg}{dx}\big|_{x_0}$	$g(x_0)$	$g(x)$	Analytical Solution $g(x) = x^3 + 1$	Error	Error %
8.5	216.75	562.000	670.375	615.125	55.250	8.98
9.0	243.00	670.375	791.875	730.000	61.875	8.48
9.5	270.75	791.875	927.250	858.375	68.875	8.02
10.0	300.00	927.250	1077.250	1001.000	76.250	7.62
10.5	330.75	1077.250	1242.625	1158.625	84.000	7.25
11.0	363.00	1242.625	1424.125	1332.000	92.125	6.92
11.5	396.75	1424.125	1622.500	1521.875	100.625	6.61
12.0	432.00	1622.500	1838.500	1729.000	109.500	6.33
12.5	468.75	1838.500	2072.875	1954.125	118.750	6.08
13.0	507.00	2072.875	2326.375	2198.000	128.375	5.84
13.5	546.75	2326.375	2599.750	2461.375	138.375	5.62
14.0	588.00	2599.750	2893.750	2745.000	148.750	5.42
14.5	630.75	2893.750	3209.125	3049.625	159.500	5.23
15.0	675.00	3209.125	3546.625	3376.000	170.625	5.05

To illustrate how to build the Excel table, take second row as example, $x_0 = 0.5$. The slope at x_0 is given by:

$$\frac{dg}{dx}\bigg|_{x_0} = 3x_0^2 = 3 \times 0.5^2 = 0.75 \qquad (2.25)$$

The value of $g(x_0)$ is the value of $g(x)$ in previous row when $x_0 = 0.0$, i.e., $g(x_0) = 1.0$. Hence, the value of $g(x_0 + h)$ estimated at $x_0 = 0.5$ is given by:

$$g(x) = g(x_0) + h\frac{dg}{dx}\bigg|_{x_0} = 0.75 + 0.5 \times 1.0 = 1.375 \qquad (2.26)$$

To determine the error in numerical integration, in this example, the true value of $g(0.5)$ can be found by substituting into the analytical solution:

$$g(x) = x^3 + 1 = 0.5^3 + 1 = 1.125 \qquad (2.27)$$

Hence, the error is $1.375 - 1.125 = 0.25$. In terms of error percentage, error% $= 0.25/1.125 = 22.22\%$. Other rows are computed similarly.

TABLE 2.5 Numerical Solution With Euler's Method at Interval 0.2
(Showing Whole Number Rows Only)

x_0	$\frac{dg}{dx}\big|_{x_0}$	$g(x_0)$	$g(x)$	Analytical Solution $g(x) = x^3 + 1$	Error	Error %
0.0	0.00	1.000	1.000	1.0	0.0	0
1.0	3.00	1.720	2.320	2.000	0.320	16.00
2.0	12.00	7.840	10.240	9.000	1.24	13.78
3.0	27.00	25.360	30.760	28.000	2.760	9.86
4.0	48.00	60.280	69.880	65.000	4.880	7.51
5.0	75.00	118.600	133.600	126.000	7.600	6.03
6.0	108.00	206.320	227.920	217.000	10.920	5.03
7.0	147.00	329.440	358.840	344.000	14.840	4.31
8.0	192.00	493.960	532.360	513.000	19.360	3.77
9.0	243.00	705.880	754.480	730.000	24.480	3.35
10.0	300.00	971.200	1031.200	1001.000	30.200	3.02
11.0	363.00	1295.920	1368.520	1332.000	36.520	2.74
12.0	432.00	1686.040	1772.440	1729.000	43.440	2.51
13.0	507.00	2147.560	2248.960	2198.000	50.960	2.32
14.0	588.00	2686.480	2804.080	2745.000	59.080	2.15
15.0	675.00	3308.800	3443.800	3376.000	67.800	2.01

Table 2.5 presents the same function but integrated numerically with a smaller step (0.2). Note that to save printed space, only whole number x values are listed. The intermediate rows are skipped but readers should be able to re-construct the whole table using the method described so far.

The results in Tables 2.4 and 2.5 are plotted in Figs. 2.6 and 2.7, respectively.

2.3.2 Modified Euler's Method

It can be seen from Fig. 2.6 that the error of Euler's method is actually quite high. This does not normally fall within the acceptable range, even the integration interval is reduced. The modified Euler's method aims to improve the accuracy. It is computationally the same as the Euler's method but instead of using the slope of the function at the start of the interval only, it

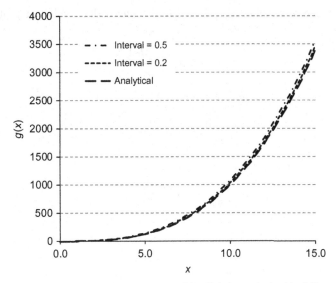

FIGURE 2.6 Comparison of responses computed by Euler's method with different integral intervals.

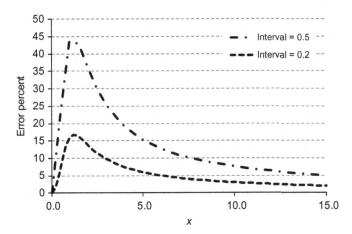

FIGURE 2.7 Graphical representation of error percentage of Euler's method.

uses an average slope for computing the step increment. First, the slope at the start of the interval x_0 is computed as:

$$\frac{dg}{dx}\bigg|_{x_0} = \frac{dg}{dx}[x_0, g(x_0)] \tag{2.28}$$

Using a new g value at $x_1 = x_0 + h$, an intermediate value of g at x_1 is calculated.

$$g_{x_1} = g(x_0) + h\frac{dg}{dx}\bigg|_{x_0} \tag{2.29}$$

Then, the new slope at x_1 is computed.

$$\frac{dg}{dx}\bigg|_{x_1} = \frac{dg}{dx}[x_1, g(x_1)] \tag{2.30}$$

The final value of g at x_1 is estimated as:

$$g(x_1) = g(x_0) + \frac{h}{2}\left(\frac{dg}{dx}\bigg|_{x_0} + \frac{dg}{dx}\bigg|_{x_1}\right) \tag{2.31}$$

To illustrate the effect of modified Euler's method on accuracy, we use a more complicated example for which the analytical solution is not immediately obvious. The initial value is $g(1) = 1$.

$$\frac{dg}{dx} = x\sqrt{g} \tag{2.32}$$

It is left to the reader to prove that the analytical solution for Eq. (2.32) is:

$$g(x) = \left(\frac{x^2 + 3}{4}\right)^2 \tag{2.33}$$

Table 2.6 presents the results for an integration interval of 0.1 using modified Euler's method. Similar to Table 2.3, the value of functions should be worked out from left to right. Table 2.7 presents the results of using Euler's method with the same integration interval of 0.1 and 0.02.

The results are compared in Fig. 2.8 and the errors are compared in Fig. 2.9.

The modified Euler's method is more accurate initially but the accuracy is getting worse as the integration continues. The choice of which numerical integration method for an application depends on what outcome is actually needed.

2.3.3 Runge–Kutta Method Second Order

A more general set of methods that is commonly used in numerical integration of differential equations is the Runge–Kutta methods. The set of methods can be applied with second order, third order, and fourth order. The order indicates how the incremental value within the interval is computed in conjunction with the division of interval. It is obvious that complexity of the method increases as the order goes up.

TABLE 2.6 Modified Euler's Method

x_0	$g(x_0)$	$\frac{dg}{dx}\big|_{x_0}$	Intermediate $g(x_1)$	$\frac{dg}{dx}\big|_{x_1}$	$\frac{1}{2}\left(\frac{dg}{dx}\big|_{x_0} + \frac{dg}{dx}\big|_{x_1}\right)$	Final $g(x_1)$	Analytical Solution $g(x) = \left(\frac{x^2+3}{4}\right)^2$	Error	Error%
1.0	1.000	1.00	1.10	1.10	1.05	1.11	1.000	0.000	0.00
1.1	1.105	1.16	1.22	1.21	1.18	1.22	1.108	−0.003	−0.25
1.2	1.223	1.33	1.36	1.34	1.33	1.36	1.232	−0.009	−0.70
1.3	1.357	1.51	1.51	1.48	1.50	1.51	1.375	−0.018	−1.31
1.4	1.506	1.72	1.68	1.64	1.68	1.67	1.538	−0.031	−2.03
1.5	1.674	1.94	1.87	1.81	1.88	1.86	1.723	−0.048	−2.81
1.6	1.862	2.18	2.08	2.01	2.10	2.07	1.932	−0.070	−3.63
1.7	2.072	2.45	2.32	2.23	2.34	2.31	2.168	−0.097	−4.46
1.8	2.305	2.73	2.58	2.47	2.60	2.57	2.434	−0.128	−5.27
1.9	2.566	3.04	2.87	2.74	2.89	2.85	2.731	−0.165	−6.05
2.0	2.855	3.38	3.19	3.04	3.21	3.18	3.063	−0.208	−6.79
2.1	3.176	3.74	3.55	3.37	3.56	3.53	3.432	−0.256	−7.46
2.2	3.531	4.13	3.94	3.73	3.93	3.92	3.842	−0.310	−8.08
2.3	3.925	4.56	4.38	4.14	4.35	4.36	4.295	−0.371	−8.63
2.4	4.359	5.01	4.86	4.58	4.80	4.84	4.796	−0.437	−9.11
2.5	4.839	5.50	5.39	5.07	5.29	5.37	5.348	−0.509	−9.51

(Continued)

TABLE 2.6 (Continued)

x_0	$g(x_0)$	$\frac{dg}{dx}\big|_{x_0}$	Intermediate $g(x_1)$	$\frac{dg}{dx}\big|_{x_1}$	$\frac{1}{2}\left(\frac{dg}{dx}\big|_{x_0} + \frac{dg}{dx}\big|_{x_1}\right)$	Final $g(x_1)$	Analytical Solution $g(x) = \left(\frac{x^2+3}{4}\right)^2$	Error	Error%
2.6	5.368	6.02	5.97	5.61	5.82	5.95	5.954	−0.586	−9.84
2.7	5.949	6.59	6.61	6.21	6.40	6.59	6.618	−0.668	−10.10
2.8	6.589	7.19	7.31	6.86	7.02	7.29	7.344	−0.755	−10.28
2.9	7.291	7.83	8.07	7.57	7.70	8.06	8.137	−0.845	−10.39
3.0	8.061	8.52	8.91	8.35	8.44	8.90	9.000	−0.939	−10.43
3.1	8.905	9.25	9.83	9.21	9.23	9.83	9.938	−1.033	−10.40
3.2	9.828	10.03	10.83	10.14	10.09	10.84	10.956	−1.128	−10.30
3.3	10.837	10.86	11.92	11.17	11.01	11.94	12.058	−1.221	−10.13
3.4	11.938	11.75	13.11	12.28	12.01	13.14	13.250	−1.311	−9.90
3.5	13.140	12.69	14.41	13.50	13.09	14.45	14.535	−1.395	−9.60
3.6	14.449	13.68	15.82	14.82	14.25	15.87	15.920	−1.471	−9.24
3.7	15.874	14.74	17.35	16.26	15.50	17.42	17.410	−1.536	−8.82
3.8	17.424	15.86	19.01	17.82	16.84	19.11	19.010	−1.586	−8.34
3.9	19.108	17.05	20.81	19.52	18.28	20.94	20.725	−1.617	−7.80
4.0	20.937	18.30	22.77	21.36	19.83	22.92	22.563	−1.626	−7.21

TABLE 2.7 Numerical Integration of the Same Function With Euler's Method (for Intervals 0.1 and 0.02)

Increment	0.1						0.02				
x_0	$g(x_0)$	$\frac{dg}{dx}\big|_{x_0}$	$g(x)$	Analytical Solution $g(x)=\left(\frac{x^2+3}{4}\right)^2$	Error	Error %	$g(x_0)$	$\frac{dg}{dx}\big|_{x_0}$	$g(x)$	Error	Error %
1.0	1.000	1.00	1.100	1.000	0.000	0.00	1.000	1.00	1.020	0.000	0.00
1.1	1.100	1.15	1.215	1.108	− 0.008	− 0.70	1.106	1.16	1.129	− 0.002	− 0.14
1.2	1.215	1.32	1.348	1.232	− 0.017	− 1.36	1.229	1.33	1.255	− 0.003	− 0.28
1.3	1.348	1.51	1.499	1.375	− 0.027	− 1.97	1.369	1.52	1.400	− 0.006	− 0.41
1.4	1.499	1.71	1.670	1.538	− 0.039	− 2.54	1.530	1.73	1.564	− 0.008	− 0.53
1.5	1.670	1.94	1.864	1.723	− 0.053	− 3.06	1.712	1.96	1.751	− 0.011	− 0.63
1.6	1.864	2.18	2.082	1.932	− 0.068	− 3.53	1.918	2.22	1.962	− 0.014	− 0.73
1.7	2.082	2.45	2.328	2.168	− 0.086	− 3.97	2.150	2.49	2.200	− 0.018	− 0.82
1.8	2.328	2.75	2.602	2.434	− 0.106	− 4.36	2.412	2.80	2.467	− 0.022	− 0.91
1.9	2.602	3.06	2.909	2.731	− 0.129	− 4.71	2.704	3.12	2.767	− 0.027	− 0.98
2.0	2.909	3.41	3.250	3.063	− 0.154	− 5.02	3.031	3.48	3.100	− 0.032	− 1.04
2.1	3.250	3.79	3.628	3.432	− 0.182	− 5.30	3.394	3.87	3.471	− 0.038	− 1.10
2.2	3.628	4.19	4.047	3.842	− 0.213	− 5.55	3.797	4.29	3.883	− 0.044	− 1.15
2.3	4.047	4.63	4.510	4.295	− 0.248	− 5.77	4.244	4.74	4.339	− 0.051	− 1.20
2.4	4.510	5.10	5.020	4.796	− 0.286	− 5.96	4.737	5.22	4.841	− 0.059	− 1.24

(Continued)

TABLE 2.7 (Continued)

Increment	0.1						0.02				
x_0	$g(x_0)$	$\frac{dg}{dx}\big|_{x_0}$	$g(x)$	Analytical Solution $g(x) = \left(\frac{x^2+3}{4}\right)^2$	Error	Error %	$g(x_0)$	$\frac{dg}{dx}\big|_{x_0}$	$g(x)$	Error	Error %
2.5	5.020	5.60	5.580	5.348	−0.328	−6.13	5.280	5.74	5.394	−0.068	−1.27
2.6	5.580	6.14	6.194	5.954	−0.374	−6.28	5.876	6.30	6.002	−0.078	−1.30
2.7	6.194	6.72	6.866	6.618	−0.424	−6.40	6.530	6.90	6.668	−0.088	−1.33
2.8	6.866	7.34	7.600	7.344	−0.478	−6.51	7.245	7.54	7.396	−0.099	−1.35
2.9	7.600	7.99	8.399	8.137	−0.537	−6.60	8.025	8.22	8.190	−0.111	−1.37
3.0	8.399	8.69	9.269	9.000	−0.601	−6.68	8.876	8.94	9.054	−0.124	−1.38
3.1	9.269	9.44	10.212	9.938	−0.670	−6.74	9.800	9.70	9.994	−0.139	−1.40
3.2	10.212	10.23	11.235	10.956	−0.744	−6.79	10.802	10.52	11.013	−0.154	−1.41
3.3	11.235	11.06	12.341	12.058	−0.823	−6.83	11.888	11.38	12.115	−0.170	−1.41
3.4	12.341	11.94	13.536	13.250	−0.908	−6.86	13.062	12.29	13.307	−0.188	−1.42
3.5	13.536	12.88	14.823	14.535	−1.000	−6.88	14.329	13.25	14.594	−0.207	−1.42
3.6	14.823	13.86	16.209	15.920	−1.097	−6.89	15.693	14.26	15.979	−0.227	−1.42
3.7	16.209	14.90	17.699	17.410	−1.200	−6.90	17.162	15.33	17.468	−0.248	−1.42
3.8	17.699	15.99	19.298	19.010	−1.311	−6.89	18.739	16.45	19.068	−0.271	−1.42
3.9	19.298	17.13	21.011	20.725	−1.428	−6.89	20.431	17.63	20.783	−0.295	−1.42
4.0	21.011	18.34	22.844	22.563	−1.552	−6.88	22.243	18.86	22.620	−0.320	−1.42

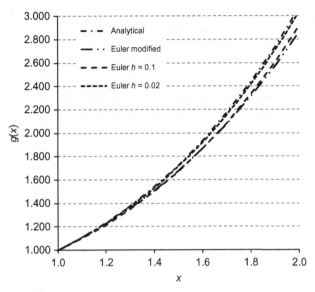

FIGURE 2.8   Graphical representation of Euler and modified Euler methods.

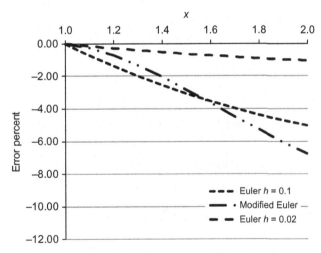

FIGURE 2.9   Graphical representation of error percentage (Euler and modified Euler methods).

Consider a function $g(x)$ with a first derivative function given by:

$$\frac{dg}{dx} = f(x, g) \tag{2.34}$$

Note that the right-hand side of the function f can include the target function g. The initial value is $g(x_0) = g_0$. The integration interval is denoted by $h = x - x_0$.

The second-order Runge–Kutta method starts with the estimation of the slope of g at the start of the interval:

$$\frac{dg}{dx}\bigg|_{x_0} = f[x_0, g(x_0)] \tag{2.35}$$

This is the first-order differential. Next, estimate the value of g at the mid-interval using the slope at the start of the interval:

$$\frac{dg}{dx}\bigg|_{x_0+0.5h} = f\left[x_0 + 0.5h, g(x_0) + 0.5h\frac{dg}{dx}\bigg|_{x_0}\right] \tag{2.36}$$

This is the second-order differential. The value of g at the end of the interval is then computed by:

$$g(x_0 + h) = g(x_0) + \frac{h}{2}\left(\frac{dg}{dx}\bigg|_{x_0} + \frac{dg}{dx}\bigg|_{x_0+0.5h}\right) \tag{2.37}$$

This process repeats by shifting the starting point of integration to $x_0 + h$.

We use Eq. (2.32) as an example. Table 2.8 presents that the tabulated values illustrating the numerical computational process.

Starting with row 1, when $x_0 = 1.0$, $g(x_0) = 1$ is given initial condition. The slope at x_0 is calculated by Eq. (2.35) as:

$$\frac{dg}{dx}\bigg|_{x_0} = f[x_0, g(x_0)] = x_0\sqrt{g(x_0)} = 1 \tag{2.38}$$

The slope at half of the integration interval is:

$$\begin{aligned}
\frac{dg}{dx}\bigg|_{x_0+0.5h} &= f\left(x_0 + 0.5h, g(x_0) + 0.5h\frac{dg}{dx}\bigg|_{x_0}\right) \\
&= (x_0 + 0.5 \times 0.1)\sqrt{g(x_0) + 0.5 \times 0.1 \times 1} \\
&= 1.076
\end{aligned} \tag{2.39}$$

The final value at the integration interval is:

$$\begin{aligned}
g(x_0 + h) &= g(x_0) + \frac{h}{2}\left(\frac{dg}{dx}\bigg|_{x_0} + \frac{dg}{dx}\bigg|_{x_0+0.5h}\right) \\
&= 1.0 + 0.5 \times 0.1(1 + 1.076) \\
&= 1.104
\end{aligned} \tag{2.40}$$

The analytical value is computed by substituting x_0 to Eq. (2.32):

$$g(x_0) = \left(\frac{x^2+3}{4}\right)^2 = \left(\frac{1^2+3}{4}\right)^2 = 1 \tag{2.41}$$

TABLE 2.8 Computational Process for Function g Using Runge–Kutta Method

x_0	$g(x_0)$	$\frac{dg}{dx}\big|_{x_0}$	$\frac{dg}{dx}\big|_{x_0+0.5h}$	$g(x_0+h)$	Analytical Solution $g(x)=\left(\frac{x^2+3}{4}\right)^2$	Error	Error%
1.00	1.000	1.000	1.076	1.104	1.000	0.000	0.0000
1.10	1.104	1.156	1.239	1.224	1.108	−0.0040	−0.3575
1.20	1.224	1.327	1.420	1.361	1.232	−0.0085	−0.6938
1.30	1.361	1.517	1.618	1.518	1.375	−0.0139	−1.0076
1.40	1.518	1.725	1.836	1.696	1.538	−0.0200	−1.2981
1.50	1.696	1.953	2.076	1.897	1.723	−0.0270	−1.5653
1.60	1.897	2.204	2.338	2.124	1.932	−0.0350	−1.8094
1.70	2.124	2.478	2.624	2.379	2.168	−0.0440	−2.0311
1.80	2.379	2.776	2.936	2.665	2.434	−0.0543	−2.2314
1.90	2.665	3.102	3.275	2.984	2.731	−0.0659	−2.4114
2.00	2.984	3.455	3.642	3.339	3.063	−0.0788	−2.5724
2.10	3.339	3.837	4.040	3.732	3.432	−0.0932	−2.7157
2.20	3.732	4.250	4.469	4.168	3.842	−0.1092	−2.8426
2.30	4.168	4.696	4.931	4.650	4.295	−0.1269	−2.9544
2.40	4.650	5.175	5.428	5.180	4.796	−0.1464	−3.0524
2.50	5.180	5.690	5.961	5.762	5.348	−0.1678	−3.1377

(Continued)

TABLE 2.8 (Continued)

x_0	$g(x_0)$	$\frac{dg}{dx}\big|_{x_0}$	$\frac{dg}{dx}\big|_{x_0+0.5h}$	$g(x_0+h)$	Analytical Solution $g(x) = \left(\frac{x^2+3}{4}\right)^2$	Error	Error%
2.60	5.762	6.241	6.531	6.401	5.954	−0.1912	−3.2115
2.70	6.401	6.831	7.141	7.100	6.618	−0.2167	−3.2750
2.80	7.100	7.461	7.791	7.862	7.344	−0.2445	−3.3289
2.90	7.862	8.131	8.483	8.693	8.137	−0.2746	−3.3744
3.00	8.693	8.845	9.218	9.596	9.000	−0.3071	−3.4121
3.10	9.596	9.603	9.999	10.576	9.938	−0.3422	−3.4429
3.20	10.576	10.407	10.826	11.638	10.956	−0.3799	−3.4675
3.30	11.638	11.258	11.701	12.786	12.058	−0.4204	−3.4865
3.40	12.786	12.157	12.626	14.025	13.250	−0.4638	−3.5005
3.50	14.025	13.107	13.602	15.360	14.535	−0.5102	−3.5100
3.60	15.360	14.109	14.630	16.797	15.920	−0.5597	−3.5154
3.70	16.797	15.164	15.712	18.341	17.410	−0.6124	−3.5173
3.80	18.341	16.274	16.850	19.997	19.010	−0.6684	−3.5159
3.90	19.997	17.440	18.045	21.772	20.725	−0.7278	−3.5117
4.00	21.772	18.664	19.298	23.670	22.563	−0.7908	−3.5050

In row 2, when $x_0 = 1.1$, $g(x_0) = 1.104$ as computed from Eq. (2.40). The slope at the current row x_0 is then calculated as:

$$\left.\frac{dg}{dx}\right|_{x_0} = f[x_0, g(x_0)] = x_0 \sqrt{g(x_0)} = 1.1 \times \sqrt{1.104} = 1.156 \qquad (2.42)$$

The slope at half of the integration interval is:

$$\left.\frac{dg}{dx}\right|_{x_0+0.5h} = (1.104 + 0.5 \times 0.1)\sqrt{1.104 + 0.5 \times 0.1 \times 1.156} = 1.239 \quad (2.43)$$

The final value at the integration interval $x_0 + h$ is:

$$\begin{aligned}
g(x_0 + h) &= g(x_0) + \frac{h}{2}\left(\left.\tfrac{dg}{dx}\right|_{x_0} + \left.\tfrac{dg}{dx}\right|_{x_0+0.5h}\right)\\
&= 1.104 + 0.5 \times 0.1(1.156 + 1.239)\\
&= 1.224
\end{aligned} \qquad (2.44)$$

The analytical value is computed by substituting x_0 to Eq. (2.33):

$$g(x) = \left(\frac{x^2+3}{4}\right)^2 = \left(\frac{1.1^2+3}{4}\right)^2 = 1.108 \qquad (2.45)$$

The other rows are computed similarly. Fig. 2.10 plots the outcomes of Runge−Kutta method. The outcomes of Euler method are also included. Fig. 2.11 compares the error in percent between Runge−Kutta and Euler methods. The Runge−Kutta method is more accurate than the Euler method.

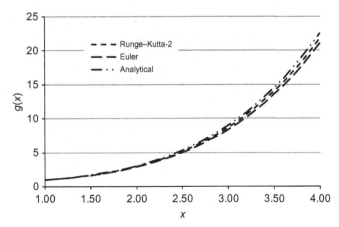

FIGURE 2.10 Function response computed by Runge−Kutta method.

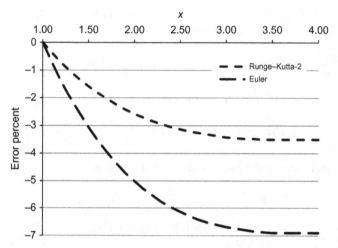

FIGURE 2.11   Comparing errors by Runge−Kutta and Euler methods.

2.3.4   Runge−Kutta Method Fourth Order

The fourth-order Runge−Kutta method starts with the first-order Taylor's expansion. The initial value of $g(x_0)$ is known. The value of g at the mid-point of the integration interval, $x_0 + h/2$, is given by:

$$g\left(x_0 + \frac{h}{2}\right) = g(x_0) + \frac{h}{2}\frac{dg}{dx}\bigg|_{x_0} = g(x_0) + \frac{h}{2}f[x_0, g(x_0)] \qquad (2.46)$$

The slope at the mid-interval is then evaluated from the g value in Eqs. (2.34) and (2.46):

$$\frac{dg}{dx}\bigg|_{x_0+\frac{h}{2},\#2} = f\left[x_0 + \frac{h}{2}, g\left(x_0 + \frac{h}{2}\right)\right] \qquad (2.47)$$

This is the second-order differential. Using Eq. (2.47), revise the estimate of g at the mid-interval, $x_0 + h/2$,

$$g\left(x_0 + \frac{h}{2}\right) = g(x_0) + \frac{h}{2}\frac{dg}{dx}\bigg|_{x_0+\frac{h}{2},\#2} \qquad (2.48)$$

Using the g value in Eq. (2.48), revise the estimate of the slope at the mid-interval, $x_0 + h/2$,

$$\frac{dg}{dx}\bigg|_{x_0+\frac{h}{2},\#3} = f\left[x_0 + \frac{h}{2}, g\left(x_0 + \frac{h}{2}\right)\right] \qquad (2.49)$$

TABLE 2.9 Computational Process for Function g for Runge–Kutta Fourth Order

x_0	$g(x_0)$	$\frac{dg}{dx}\big\|_{x_0}$	$g(x_0 + 0.5h)$	$\frac{dg}{dx}\big\|_{x_0+0.5h,\#2}$	$g(x_0 + 0.5h)$	$\frac{dg}{dx}\big\|_{x_0+0.5h,\#3}$	$g(x_0 + h)$	$\frac{dg}{dx}\big\|_{x_0+h}$	$g(x_0 + h)$	Analytical Solution $g(x) = \left(\frac{x^2+3}{4}\right)^2$	Error	Error%
1.00	1.000	1.000	1.100	1.1012	1.0551	1.0785	1.1079	1.1578	1.109	1.000	0.000	0.000
1.10	1.109	1.158	1.224	1.2725	1.1722	1.2451	1.2331	1.3326	1.234	1.108	0.0009	0.0782
1.20	1.234	1.333	1.367	1.4617	1.3071	1.4291	1.3770	1.5255	1.378	1.232	0.0020	0.1588
1.30	1.378	1.526	1.531	1.6702	1.4616	1.6321	1.5413	1.7381	1.543	1.375	0.0033	0.2402
1.40	1.543	1.739	1.716	1.8997	1.6375	1.8555	1.7281	1.9719	1.730	1.538	0.0049	0.3212
1.50	1.730	1.973	1.927	2.1516	1.8371	2.1009	1.9396	2.2283	1.941	1.723	0.0069	0.4005
1.60	1.941	2.229	2.164	2.4274	2.0627	2.3697	2.1783	2.5090	2.180	1.932	0.0092	0.4772
1.70	2.180	2.510	2.431	2.7287	2.3166	2.6636	2.4466	2.8155	2.449	2.168	0.0119	0.5507
1.80	2.449	2.817	2.730	3.0569	2.6015	2.9839	2.7471	3.1491	2.749	2.434	0.0151	0.6204
1.90	2.749	3.151	3.065	3.4136	2.9202	3.3323	3.0827	3.5115	3.085	2.731	0.0187	0.6860
2.00	3.085	3.513	3.437	3.8004	3.2754	3.7101	3.4564	3.9042	3.459	3.063	0.0229	0.7473
2.10	3.459	3.906	3.850	4.2186	3.6703	4.1190	3.8713	4.3286	3.875	3.432	0.0276	0.8042
2.20	3.875	4.330	4.308	4.6698	4.1080	4.5603	4.3305	4.7863	4.334	3.842	0.0329	0.8568
2.30	4.334	4.788	4.813	5.1555	4.5919	5.0358	4.8377	5.2787	4.842	4.295	0.0389	0.9051
2.40	4.842	5.281	5.370	5.6773	5.1255	5.5467	5.3963	5.8075	5.401	4.796	0.0455	0.9492
2.50	5.401	5.810	5.982	6.2366	5.7124	6.0947	6.0100	6.3740	6.015	5.348	0.0529	0.9894

(Continued)

TABLE 2.9 (Continued)

x_0	$g(x_0)$	$\frac{dg}{dx}\big|_{x_0}$	$g(x_0+0.5h)$	$\frac{dg}{dx}\big|_{x_0+0.5h,\#2}$	$g(x_0+0.5h)$	$\frac{dg}{dx}\big|_{x_0+0.5h,\#3}$	$g(x_0+h)$	$\frac{dg}{dx}\big|_{x_0+h}$	$g(x_0+h)$	Analytical Solution $g(x)=\left(\frac{x^2+3}{4}\right)^2$	Error	Error%
2.60	6.015	6.376	6.652	6.8349	6.3564	6.6812	6.6828	6.9798	6.688	5.954	0.0611	1.0257
2.70	6.688	6.982	7.386	7.4738	7.0615	7.3077	7.4186	7.6264	7.424	6.618	0.0701	1.0585
2.80	7.424	7.629	8.187	8.1546	7.8317	7.9758	8.2216	8.3153	8.227	7.344	0.0799	1.0880
2.90	8.227	8.318	9.059	8.8791	8.6714	8.6869	9.0961	9.0479	9.102	8.137	0.0907	1.1143
3.00	9.102	9.051	10.007	9.6486	9.5848	9.4426	10.0467	9.8259	10.053	9.000	0.1024	1.1377
3.10	10.053	9.829	11.036	10.4646	10.5766	10.2443	11.0778	10.6507	11.085	9.938	0.1151	1.1584
3.20	11.085	10.654	12.150	11.3287	11.6514	11.0936	12.1944	11.5237	12.202	10.956	0.1289	1.1766
3.30	12.202	11.527	13.355	12.2423	12.8142	11.9920	13.4012	12.4466	13.409	12.058	0.1438	1.1925
3.40	13.409	12.450	14.654	13.2070	14.0698	12.9408	14.7035	13.4208	14.712	13.250	0.1598	1.2063
3.50	14.712	13.425	16.055	14.2242	15.4234	13.9418	16.1064	14.4478	16.116	14.535	0.1771	1.2181
3.60	16.116	14.452	17.561	15.2956	16.8804	14.9963	17.6152	15.5291	17.625	15.920	0.1955	1.2281
3.70	17.625	15.533	19.178	16.4224	18.4462	16.1059	19.2356	16.6662	19.246	17.410	0.2153	1.2365
3.80	19.246	16.671	20.913	17.6063	20.1263	17.2720	20.9732	17.8606	20.984	19.010	0.2364	1.2434
3.90	20.984	17.865	22.771	18.8488	21.9265	18.4962	22.8337	19.1139	22.845	20.725	0.2588	1.2489
4.00	22.845	19.119	24.757	20.1514	23.8528	19.7799	24.8232	20.4274	24.835	22.563	0.2828	1.2532

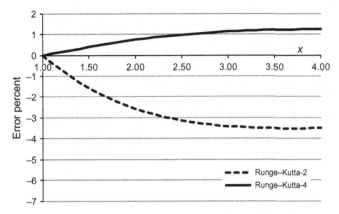

FIGURE 2.12   Comparing errors by the two Runge−Kutta methods.

This is the third-order differential. Using Eq. (2.49), estimate of the end of interval g at $x = x_0 + h$,

$$g(x_0 + h) = g(x_0) + h\frac{dg}{dx}\bigg|_{x_0 + \frac{h}{2}, \#3} \tag{2.50}$$

Using the g value in Eq. (2.50), estimate of the slope at the end interval, $x_0 + h$,

$$\frac{dg}{dx}\bigg|_{x_0 + h} = f[x_0 + h, g(x_0 + h)] \tag{2.51}$$

This is the fourth-order differential. Finally, the g value at the end of this integration interval is given by:

$$g(x_0 + h) = g(x_0) + \frac{h}{6}\left(\frac{dg}{dx}\bigg|_{x_0} + 2\frac{dg}{dx}\bigg|_{x_0 + \frac{h}{2}, \#2} + 2\frac{dg}{dx}\bigg|_{x_0 + \frac{h}{2}, \#3} + \frac{dg}{dx}\bigg|_{x_0 + h}\right) \tag{2.52}$$

We use Eq. (2.32) again as an example. Table 2.9 presents that the working table illustrating the numerical computational process. Fig. 2.11 compares the errors by the two Runge−Kutta methods.

Fig. 2.12 compares the error percent of the two Runge−Kutta methods.

Chapter 3

Wind Power and Aerodynamics Systems

3.1 LIQUID FLOW SYSTEMS

Wind is a form of solar energy that are caused by many environmental factors such as the sun's heating of the atmosphere, rotation of the earth, and irregularities of the earth's surface. Wind turbines convert the kinetic energy in the wind into mechanical power. This can then be used for specific tasks but more commonly, a generator is used to convert the mechanical power into electricity. These are large rotating machines in which its components, such as the blades and blade joints, are subjected to high wind forces. This force can limit the life of the blades causing them to fail, subjecting the wind turbine to major repairs.

3.2 BASIC KNOWLEDGE AND TERMINOLOGY OF WIND TURBINE

The analysis of loading and strength of a blade is important during normal operation, as the blade's exterior shape and structure is essential for the wind turbine to have sufficient strength, stiffness, and stability. To understand the blade strength, analysis of aerodynamic loading must be carried out. The most commonly used methods for calculation the aerodynamic loading acting on the blade include blade element momentum (BEM) theory, computational fluid dynamics (CFD), and wind tunnel testing among others. The BEM theory is the common principle to analyze the aerodynamic loading, which is based on the combination of momentum theory and blade element theory. To simplify the analysis, some key assumptions are made in the method:

- Flow is regarded as incompressible fluid
- Sudden drop in pressure occurs through the rotor disk
- Flow velocity is continuous and varies smoothly from upstream to downstream
- Rotating flows in the wake of rotor is ignored
- Flow is considered as steady state
- Mass flow rate of wind is constant

Demystifying Numerical Models. DOI: https://doi.org/10.1016/B978-0-08-100975-8.00003-5

3.3 BLADE ELEMENT THEORY

BEM theory equates two methods of examining how a wind turbine operates. The first method is to use a momentum balance on a rotating annular stream tube passing through a turbine. The second is to examine the forces generated by the airfoil lift and drag coefficients at various sections along the blade. These two methods then give a series of equations that can be solved iteratively.

3.3.1 Axial Force and Momentum Change

Consider the stream tube and the actuator disc as shown in Fig. 3.1. Four stations are shown in the diagram: (1) some way upstream of the turbine, (2) just before the blades, (3) just after the blades, and (4) some way downstream of the blades. Between (2) and (3) energy is extracted from the wind and there is a change in pressure as a result.

To estimate the kinetic energy loss due to momentum changes, some assumptions are taken in the analysis:

- Assume far field pressure is constant
- No friction loss along the tube, expect across the actuator disc
- The fluid flow obeys the Bernoulli's equation

Based on the aforementioned assumptions, the volume flow rate remains constant within the stream tube. However, due to reduction of velocity before and after the wind turbine, the upstream of the stream tube has a smaller area than the downstream. The cross-section area is expanded due to velocity reduction (i.e., fundamental of mass conservation). The mass flow rate is therefore considered to be constant:

$$\rho A_\infty U_\infty = \rho A_d U_d = \rho A_W U_W \tag{3.1}$$

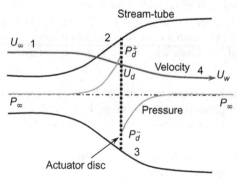

FIGURE 3.1 Actuator and stream tube stations for a wind turbine.

Notice that the density of the air remains constant in most of the cases. In another word, only the area and the velocity of the stream tube are the variables in the equation. To simplifier the equation, we introduces a factor called axial flow induction factor (or referred as the inflow factor in some other textbook):

$$a = \frac{U_\infty - U_d}{U_\infty} \quad \text{or} \quad U_d = U_\infty(1 - a) \tag{3.2}$$

The axial forces acting on the wind turbine is resulted from the rate of change in momentum in axial direction. Theoretically, the rate of change in momentum equals to the overall change in velocity (i.e., velocity entering and exiting the stream tube) time the mass flow rate passing through the wind turbine which is given by:

$$(U_\infty - U_W)\rho A_d U_d \tag{3.3}$$

As shown in Fig. 3.1, the axial force acting on the wind turbine occurs in the form of the pressure difference across the front and the back of actuator disc. Such pressure difference could also be viewed as the force induced by the wind pressure. The pressure difference across the disc can then be equated to the rate of change in momentum as follow:

$$(P_d^+ - P_d^-)A_d = (U_\infty - U_W)\rho A_d U_\infty(1 - a) \tag{3.4}$$

To estimate the power output of the wind turbine, it is essential to evaluate the axial force from the above equation. From the above equation, one could notice that the axial force is directly proportional to the pressure difference acting on the turbine. The Bernoulli's equation is used to estimate the pressure difference:

$$\frac{1}{2}\rho_\infty U_\infty^2 + P_\infty + \rho_\infty g h_\infty = \frac{1}{2}\rho_d U_d^2 + P_d^+ + \rho_d g h_d \tag{3.5}$$

Similarly, for the downstream (i.e., right-hand side of the disc as shown in Fig. 3.1), the pressure difference can be expressed as follow:

$$\frac{1}{2}\rho_W U_W^2 + P_\infty + \rho_W g h_W = \frac{1}{2}\rho_d U_d^2 + P_d^- + \rho_d g h_d \tag{3.6}$$

Assuming the flow is incompressible and the change in potential energy is negligible:

$$\rho_\infty = \rho_d = \rho_W \quad \text{and} \quad h_\infty = h_d = h_W \tag{3.7}$$

Subtracting Eq. (3.7) from Eq. (3.6) gives:

$$(P_d^+ - P_d^-) = \frac{1}{2}\rho(U_\infty^2 - U_W^2) \tag{3.8}$$

Substitute to Eq. (3.4) gives:

$$\frac{1}{2}\rho(U_\infty^2 - U_W^2)A_d = (U_\infty - U_W)\rho A_d U_\infty(1 - a) \qquad (3.9)$$

After some simple algebraic procedures, one could also obtain the following equation:

$$U_W = (1 - 2a)U_\infty \qquad (3.10)$$

From the physical viewpoint, the above equation presents the idealization of the velocity loss within the stream tube. Half of the speed loss takes place in the upstream (i.e., from the free stream velocity to the disc velocity) whereas the other half takes place in the downstream.

The power generation caused by the axial force equals to the work done by the pressure done times the velocity of the wind at the actuator disc. The pressure force is given by:

$$F = (P_d^+ - P_d^-)A_d = (U_\infty - U_W)\rho A_d U_\infty(1 - a) \qquad (3.11)$$

with Eq. (3.10), the pressure force can then be expressed as:

$$\begin{aligned} F &= [U_\infty - U_\infty(1 - 2a)]\rho A_d U_\infty(1 - a) \\ F &= 2\rho A_d U_\infty^2 a(1 - a) \end{aligned} \qquad (3.12)$$

The pressure force acting on a finite element of the wind turbine is therefore given by:

$$dF_x = 4a(1 - a)\rho U_\infty^2 \pi r\, dr \qquad (3.13)$$

3.3.2 Rotating Angular Momentum

In practice, most of the wind turbine employ a rotor with blades rotating with an angular speed on an axis parallel to the wind direction. In terms of momentum exchange, axial momentum loss is converted into rotating angular momentum through the exertion of torque and thrust by the wind. While torque is generated on the blade of the wind turbine, the reaction of the torque also causes the air to rotate in a direction opposite to that of the rotor. The rotating motion of the air compensate the static pressure loss of the wind. Fig. 3.2 illustrates the change in the velocity and momentum of wind passing through the wind turbine.

From the physical viewpoint, the above equation presents the idealization of the velocity loss within the stream tube. Half of the speed loss takes place in the upstream (i.e., from the free stream velocity to the disc velocity) whereas the other half takes place in the downstream. The blades of the wind turbine alter the wind direction and induce a rotating motion downstream. The induced tangential velocity varies according to the radial

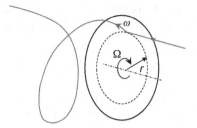

FIGURE 3.2   Change in wind velocity and momentum across wind turbine.

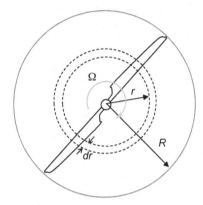

FIGURE 3.3   Illustrated annular ring on the rotor disc.

location along the blade resulting different angular momentum exchange and force generation on the blade. Obviously, the torque generate on the blade is proportional to the rate of change in angular momentum. A detail analysis of angular momentum change for the whole blade is essential for estimating the torque and power output of the wind turbine.

Analysis of the angular momentum is carried out on an annular ring of the rotor disc with a finite width in radial direction of dr. Fig. 3.3 depicts the visualization of the annular ring as part of the rotor disc for angular momentum analysis. The mass flow rate of wind passing through the annular ring is therefore given by:

$$\dot{m} = \rho U_\infty (1 - a) 2\pi r\, dr \qquad (3.14)$$

Assuming the wind leaving the blade with an angular velocity of ω, torque acting on the annular ring equals to the rate of change in angular momentum where is given by:

$$dT = \rho U_\infty (1 - a) 2\pi r\, dr (\omega r) r \qquad (3.15)$$

Similar to the axial momentum analysis, we introduce an angular induction factor to simplify the equation. The angular induction factor is defined as:

$$a' = \frac{\omega}{2\Omega} \tag{3.16}$$

where Ω is the angular rotational speed of the wind turbine. The torque acting on the annular ring can thereby be written as follow:

$$dT = 4\rho U_\infty (1-a)\frac{\omega}{2\Omega}\Omega\pi r^3 \, dr$$

$$dT = 4a'(1-a)\rho U_\infty \Omega\pi r^3 k \, dr \tag{3.17}$$

The above equation, in combination with Eq. (3.13), forms two fundamental equations for calculating the pressure and tangential forces acting on an annular element of the turbine due to axial and angular momentum changes. In fact, it will become more trivial that these two equations form a closing loop of relating the power output of the turbine with the axial and angular induction due to the turbine motion.

3.3.3 Blade Element and Relative Velocity

A closer examination of the above derivation reveal that Eq. (3.17) only estimate the torque acting an annular ring of the rotor disc with a finite width in radial direction of dr. In short, at least in the first principle, it requires integration to evaluate the total torque acting on the whole blade (or all blades of the wind turbine). Nonetheless, as discussed in the previous section, blades of a practical wind turbine usually have a varying chord length, twisting angle, and chamber height along the radial direction. Integrating the torque for the whole blade is obviously sophisticated and almost impractical. To simplify the problem, numerical integration could be employed dividing the whole blade into many elements (i.e., referred as blade element) and evaluating the forces acting on each element in finite sections.

Considering a typical blade divided into N number of element as shown in Fig. 3.4. As depicted, each of the blade element represents different radial location as well as chord length, twisting angle, and chamber height. It is also worthwhile to notice that each element will also subject to different rotational speed (i.e., Ωr) owing to its radial location. The difference in rotational speed could pose significant impact on the forces generated on the blade. The aerodynamic impact on each blade element is related to the relative velocity arriving at the leading edge of the blade. To analysis its impact, it is however essential to grasp a clear understanding of relative velocity concept.

Fig. 3.4 shows a typical wind turbine with two blades rotating around an axis (i.e., the axis is pointing into the book) in the clockwise direction. To

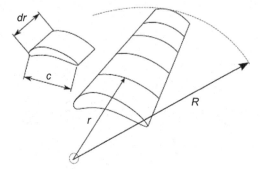

FIGURE 3.4 Blade divided into a number of elements [1].

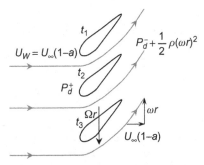

FIGURE 3.5 Snapshots of blade tip moving downwards.

visualize the relative velocity frame, one could imagine putting a viewpoint on the right of the figure (i.e., Fig. 3.3) focusing on the sectional view of tip of the blade (just like any blade element). At a given series of time frame, on the tangential line of the blade circumference, one could visualize that the blade tip is moving downwards in analog to linear motion. Fig. 3.5 shows the visualization of the motion of the blade tip moving downward within a series of time frame (i.e., t_1 to t_3). As depicted, the blade tip is moving at the tangential velocity (i.e., Ωr); forming a blade row cutting through the horizontal wind. The incoming wind velocity is U_∞ and the velocity of the wind passing through the blade is of U_d (for details refer to Section 3.3.1). As discussed above, action of blade induces a rotating motion to the wind downstream where the rotational speed is notated as ω.

As air at the inlet of the blade is not rotating, at the exit of the blade row, the flow rotates with the corresponding tangential velocity of ωr. The average rotational flow downstream is therefore estimated as $\omega/2$. Considering the blade rotational speed (which is in opposite rotation direction), the average relative velocity that the blade is experienced is given by:

$$\Omega r + \frac{\omega r}{2} \tag{3.18}$$

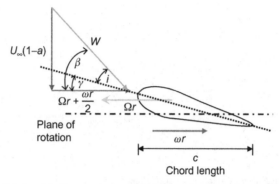

FIGURE 3.6 Relative velocity components at the blade.

Expressing the equation in term of the angular induction factor gives:

$$\Omega r + \frac{\omega r}{2} = \Omega r(1 + a') \qquad (3.19)$$

A more detail illustration of the above relative velocity concept is shown in Fig. 3.6.

Combining the relative rotational velocity with the passing wind speed, as shown in Fig. 3.6, it forms a relative wind speed (i.e., notated as W) that determine the aerodynamics characteristics and performance of the blade subject to. According to Eq. (3.19), the resultant relative wind speed is given by:

$$W = \sqrt{U_\infty^2 (1-a)^2 + \Omega^2 r^2 (1+a')^2} \qquad (3.20)$$

The relative wind direction can be calculated from the following relationship:

$$\tan \beta = \frac{U_\infty(1 - a)}{\Omega r(1 + a')} \qquad (3.21)$$

Define the local tip speed ratio as:

$$\lambda_r = \frac{\Omega r}{U_\infty} \qquad (3.22)$$

The relative wind speed and direction could be further simplified as follow:

$$W = \frac{U_\infty(1 - a)}{\sin \beta}$$

$$\tan \beta = \frac{(1 - a)}{\lambda_r(1 + a')} \qquad (3.23)$$

It is worthwhile to notice that the local tip speed ratio should not be confused with the overall tip speed ratio of the wind turbine. The local tip speed ratio indicates the ratio of the tangential velocity at the given radial position to the incoming wind speed. In other word, the local tip speed ratio is variable depending on the radial position along the blade.

Through the illustration of relative velocity concept, one could now observe that the blade element is subject to the relative wind speed W instead of the incoming wind speed. The relative wind speed however is determined by the incoming wind speed and the rotational speed of the blade. Obviously, from the physical viewpoint, the rotational speed of the blade is driven by the aerodynamic forces induced by the passing relative wind speed. Both relative wind speed and rotational speed are coupled and balanced each other through the aerodynamic forces acting blade in practical system. This is also the main reason that numerical iterative procedures are required to resolve the forces, rotational speed, and power output via trial-and-error approach.

3.3.4   Aerodynamic Lift and Drag Forces

Based on the relative velocity concept, it is now possible to relate the aerodynamic performance of the moving blade to the available lift and drag coefficient data that largely measured in wind tunnel, where the airfoil was tested in stationary. Fig. 3.7 shows the definition of lift and drag forces acting on the moving blade. As depicted, the drag force is defined as the aerodynamic force acting parallel to the chord of the blade element while the lift force is perpendicular to the chord.

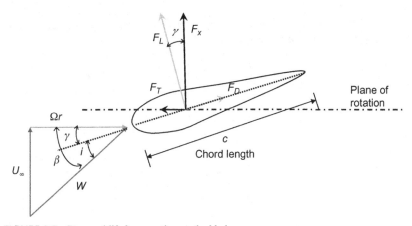

FIGURE 3.7   Drag and lift forces acting on the blade.

The drag force acting on the blade is parallel to the direction of resultant relative velocity which can be expressed as:

$$dD = C_D \frac{1}{2} \rho W^2 c \; dr \tag{3.24}$$

Similarly, the lift force acting on the perpendicular direction is given as:

$$dL = C_L \frac{1}{2} \rho W^2 c \; dr \tag{3.25}$$

where C_D and C_L are the drag and lift coefficients available in the wind tunnel data.

The generated lift and drag forces are however misalign with the plane of rotation. In practice, it would be more convenient to relate all the force in terms of trust and torque acting on the blade. For each blade, the trust on the blade with respect to the rotational axis (i.e., force in the axial direction) is given as:

$$F_x = dL \cos \beta + dD \sin \beta \tag{3.26}$$

Meanwhile, the torque force acting on the tangential direction is expressed as:

$$F_T = dL \sin \beta - dD \cos \beta \tag{3.27}$$

The above two equations could be more useful by expressing the thrust and torque in terms of relative wind direction and induction factor as follow:

$$F_x = \frac{1}{2} \rho \frac{U_\infty^2 (1-a)^2}{\sin^2 \beta} (C_L \cos \beta + C_D \sin \beta) c \; dr$$

$$F_T = \frac{1}{2} \rho \frac{U_\infty^2 (1-a)^2}{\sin^2 \beta} (C_L \sin \beta - C_D \cos \beta) c \; dr \tag{3.28}$$

Consider there are N number of blades on the turbine, the total generated thrust and torque are then given by:

$$dF_x = \frac{1}{2} N \rho \frac{U_\infty^2 (1-a)^2}{\sin^2 \beta} (C_L \cos \beta + C_D \sin \beta) c \; dr$$

$$dT = \frac{1}{2} N \rho \frac{U_\infty^2 (1-a)^2}{\sin^2 \beta} (C_L \sin \beta - C_D \cos \beta) c \; dr \tag{3.29}$$

Recalling from the previous session, the torque and trust could also be derived from the axial and angular momentum theory as follow:

$$dF_x = 4a(1-a)\rho U_\infty^2 \pi r \; dr$$
$$dT = 4a'(1-a)\rho U_\infty \Omega \pi r^3 \; dr \tag{3.30}$$

Combining the above four equations yields:

$$\frac{a}{1-a} = \frac{\sigma'(C_L \cos \beta + C_D \sin \beta)}{4 \sin^2 \beta}$$

$$\frac{a'}{1-a} = \frac{\sigma'(C_L \sin \beta - C_D \cos \beta)}{4\lambda_r \sin^2 \beta} \qquad (3.31)$$

where σ' is the solidity of the given wind turbine which is given as:

$$\sigma' = \frac{Nc}{2\pi r} \qquad (3.32)$$

Consider that the relative wind direction is given as:

$$\tan \beta = \frac{(1-a)}{\lambda_r(1+a')} \qquad (3.33)$$

The above three equations form a set of system equations for the given blade element of a wind turbine. Mathematically, for a given local tip speed ratio, there are more than three unknown within the three equations (i.e., a, a', C_D, C_L, and β). Fortunately, the lift and drag coefficients could be evaluated base on the relative wind direction β and the vastly available wind tunnel database.

In the past decades, many series of wind tunnel experiment have been carried out by many large research consortium (e.g., NASA, NREL, etc.) aiming to test a variety of airfoils and its aerodynamic performance at different flow situations. These data are mostly available online or within literatures in the public domain.

As discussed before, most of the wind tunnel data were tested under the situation where the tested airfoil was mounted stationary with the testing wind speed controlled by the fan. Therefore, most of the wind tunnel data present the measured lift and drag coefficients against various angle of attack. Fig. 3.8 shows the measured lift and drag coefficients for an airfoil (S809) at various angles of attack. Most of the wind tunnel data consist of more than one testing wind speed. To make the data scalable and dimensionless, the testing wind speed is expressed as Reynolds number which is defined as:

$$\text{Re} = \frac{\rho U_\infty c}{\mu} \qquad (3.34)$$

where ρ and μ are the density and dynamics viscosity of the wind air.

As depicted, for the selected airfoil (S809), large amount of lift force could be generated at a relatively low value of angle of attack (i.e., $<5°$). The two Reynolds number corresponding to two wind speeds exhibit considerably insignificant impact on the resultant lift force at low angle of attack. Above the low angle of attack, the lift coefficient appears a

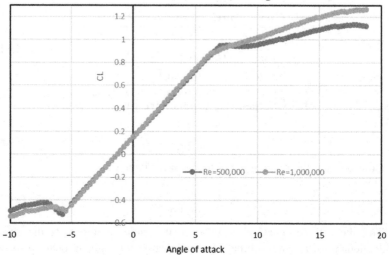

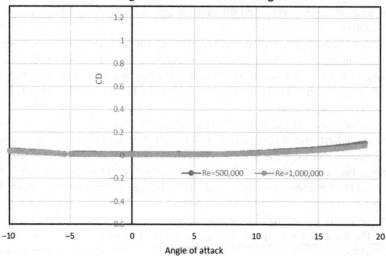

FIGURE 3.8 Variation of the lift and drag coefficients at various angle of attack of the selected airfoil (S809) [2].

relatively slight increase until the stall condition occurs at roughly 19°. Comparing to the lift coefficient, the drag coefficient remains mostly constant close to zero. This implies that the selected airfoil has a very lift to drag ratio which is favorable for wind energy harvesting application.

From Fig. 3.8, one could observe that the lift and drag coefficients for a given airfoil could be readily determined using the existing wind tunnel data. Therefore, the number of unknown could be reduced making it possible to resolve the solution using numerical method. Nevertheless, it is also worth noting that wind tunnel data are mostly measured at specific Reynolds number range. It is crucial to ensure that the lift and drag estimations are within the range. Extrapolating wind tunnel data is not recommended and normally embedded significant error.

3.3.5   Losses at the Tip and Overall Power Output

Before presenting the detailed procedures for solving the blade element method, it is important to understand the possible energy due to the vortex generated at the tip of the blade. In the previous session, drag and lift coefficients extracted from wind tunnel data are introduced. It is worth noting that majority these wind tunnel data were measured from two-dimensional airfoil where turbulence vortex generated at the tip as well as the characteristic of continuous twisting and tapering blade are not considered. To account for these losses, it is common to introduce a correction factor into Eq. (3.30).

The correction factor is a dimensionless number ranging from 0 to 1 characterizing the reduction of the force acting on the blade as follow:

$$Q = \frac{2}{\pi} \cos^{-1} \left[\exp \left(-\frac{N/2(1 - r/R)}{(r/R)\cos \beta} \right) \right] \tag{3.35}$$

After introducing the tip loss correction factor, the original equations become:

$$\begin{aligned} dF_x &= Q4a(1 - a)\rho U_\infty^2 \pi r \, dr \\ dT &= Q4a'(1 - a)\rho U_\infty \Omega \pi r^3 \, dr \end{aligned} \tag{3.36}$$

Similar to Eq. 3.31:

$$\begin{aligned} \frac{a}{1 - a} &= \frac{\sigma'(C_L \cos \beta + C_D \sin \beta)}{4Q \sin^2 \beta} \\ \frac{a'}{1 - a} &= \frac{\sigma'(C_L \sin \beta - C_D \cos \beta)}{4Q\lambda_r \sin^2 \beta} \end{aligned} \tag{3.37}$$

Since the correction factor is a function of the relative wind direction β, the additional consideration in the equation does not affect the validity of the system of equations.

The one last equation to complete the analysis is calculating the total power yield from each blade element and thus estimating the overall power output of the whole wind turbine. With the equation trust, the power yield from each blade element is expressed as:

$$dP = \Omega \, dT \tag{3.38}$$

In principle, the overall power output from one whole blade is given as:

$$P = \int_{r_h}^{R} dP \, dr \tag{3.39}$$

where r_h is the radius of the rotor hub.

With the above equation, one could notice that the integral representing the power yield along the whole blade along the radial direction. As discussed, most of the modern blade design has a continuous twisting and tapering configuration. The complicity of the blade geometry poses substantial difficulties in the evaluating the integral of power output. Numerical integral approach could be adopted where more example will be seen in the following sessions.

On the other hand, for the ease of comparing each blade designs, it is also common to evaluate the power coefficient of the blade which is defined as:

$$C_P = \frac{Power}{(1/2)\rho A_d U_\infty^3} = \frac{\int_{r_h}^{R} dP \, dr}{(1/2)\rho A_d U_\infty^3} \tag{3.40}$$

After some algebraic steps, the power coefficient is given as:

$$C_P = \frac{8}{\lambda^2} \int_{\lambda_h}^{\lambda} Q \lambda_r^3 a'(1-a) \left[1 - \frac{C_D}{C_L} \tan \beta \right] d\lambda_r \tag{3.41}$$

where λ_h is the local tip speed ratio at the rotor hub.

3.4 BLADE DESIGN AND SOLVING PROCEDURES

3.4.1 Overall Blade Design Procedure

Depends on the given problem or the design criteria, the iterative procedure could be slightly different due to the fact that some design parameters could be subjected to constraint and remain constant. Nonetheless, for most of the problems, the overall iterative procedures could follow the steps listed as follows:

1. Based on the given wind condition or site location, design the targeted power output of the wind turbine which could lead to determine the following design parameters:
   - Radius of the blade tip (i.e., R) and the hub (i.e., r_h)
   - Expected or operating wind speed (i.e., V)
   - Expected overall electrical and mechanical efficiencies
   - Estimated power of coefficient (i.e., Cp)

2. Design or determine the desired tip speed ratio (i.e., λ) of the wind turbine. The tip speed ratio is however limited by the noise generation at the blade tip. Rule of thumb says the noise from a wind turbine increases with the fifth power of the blade speed. Therefore, a small increment of tip speed ratio can make a large difference. Tip speed can be limited to approximately 60 m/s. Considering a typical wind farm with wind speed around 10−15 m/s, tip speed ratio is ranging from 4 to 6.

3. Select or design the number of blade N. According to practical experience [3], the number of blade could be determined from the tip speed ratio as presented in Table 3.1. Evaluate the performance of wind turbine using BEM method and modify or optimize the design as desired.

4. Choosing a suitable airfoil based on the desired tip speed ratio. For high tip speed ratio (i.e., $\lambda > 3$), a high lift airfoil is commonly adopted. In contrast, for low tip speed ratio, a curve plate or high drag plate could be adopted instead of airfoil.

5. Obtain or extract the existing drag and lift coefficient wind tunnel data for the selected airfoil. Notice that some wind turbine designs could consist of different airfoil profile in at different spans of the blade.

6. Design the operating aerodynamics condition for the given airfoil. In practical, it is commonly adopted 80% of the maximum lift value as the desired operating condition. The operating condition leads to the choice of span wise twist of the blade. Notice that the blade twist near the hub is also constrained by the structure integrity.

7. Design the chord distribution of the airfoil. It is worth nothing that designing chord distribution is complicated and subject to various engineering considerations such as aerodynamic performance, weight and mass distribution, structure strength, and integrity. However, a simplified equation for expressing the blade distribution of an ideal blade can be written as:

$$C = \frac{8\pi r \cos \beta}{3N\lambda_r} \qquad (3.42)$$

TABLE 3.1 Recommended Number of Blade Against Typical Tip Speed Ratio

Tip Speed Ratio, λ	Number of Blade, N
1	8−24
2	6−12
3	3−6
4	3−4
>4	1−3

3.4.2 Iterative Procedure for Solving the BEM

In the previous session, the final forms of the blade element thrust and torque derived by momentum balance are expressed in terms of variables such as relative wind direction, axial and angular induction factor. Since all variables are coupled each other, the set of equations cannot be solved directly using analytical method. Nevertheless, it is possible to adopt numerical iterative method for obtaining the solution by trail-and-error approach.

Depends on the given problem or the design criteria, the iterative procedure could be slightly different due to the fact that some design parameters could be subjected to constraint and remain constant. Nonetheless, for most of the problems, the overall iterative procedures could follow the steps listed as follows:

1. Based on the given wind condition or desired operating condition, determine the range of Reynolds number or flow condition which the airfoil is subject to.
2. Obtain or examine the lift and drag coefficients of the selected airfoil. This could be extracted from existing wind tunnel database or even obtained from physical test.
3. Divide the whole blade into N elements along the span wise direction. Typically, it requires 10 to 20 elements for obtaining a reliable estimation.
4. Estimate a guess value for the induction factors (i.e., a and a') and the relative wind direction (i.e., β) to start the iterative process. In theory, the guess value should be independent to the final converged result. In other words, any guess value could be used as far as the iterative could converge. Nonetheless, a reasonable value could accelerate the iterative process. The guess value could be estimated from the following equations where it zero drag and zero tip loss are assumed. This is however an initial guess value ONLY.

$$\beta = \frac{2}{3}\tan^{-1}\left(\frac{1}{\lambda_r}\right)$$

$$a = \left(1 + \frac{4\sin^2\beta}{\sigma' C_L \cos\beta}\right)^{-1} \qquad (3.43)$$

$$a' = \frac{1 - 3a}{4a - 1}$$

5. Repeating the following iterative steps until all unknown converged.
   - Calculate the beta and tip speed ratio
   - Obtain the lift and drag coefficients for the appropriate incidence angle
   - Calculate and update the a and a'
6. Evaluate the rotor performance and power output based on the converged value.

3.4.3   Power Output of a Twisted and Tapered Blad (NACA S809)

To demonstrate the detail of the iterative procedure, this session presents an example with appropriate spreadsheet template for readers. Considering the variation of local tip speed ratio and angle of attack along the blade, a twisted and tapered blade was designed to operate on a stall-regulated downwind wind turbine. The blade was developed based on the geometry of a standard NACA airfoil S809 which has a linear taper and nonlinear twist distributions along the radial direction. The cross-section profile of the selected airfoil is depicted in Fig. 3.9. The wind turbine could install two to three blades. Each of the blade has a radius of 5 m (i.e., from root to tip). The wind turbine is designed to operate at the wind speed of 7.2 m/s with the rotational speed of 72 rpm. The average temperature of the coming wind is assumed to 25°C.

Details of the blade twisting angle and corresponding length at various radial locations are listed in Table 3.2. Furthermore, together with the blade

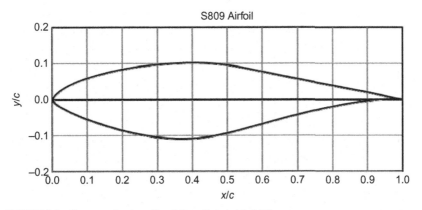

FIGURE 3.9   Cross-sectional profile of the selected airfoil [4].

TABLE 3.2 Twisting Angle and the Chord Distribution of the Blade

Radius (m)	Twisting Angle (°)	Chord Length (m)
2	3.158	0.661
2.5	2.270	0.611
3	1.647	0.560
3.5	0.481	0.510
4	− 0.525	0.460
5	− 1.975	0.359

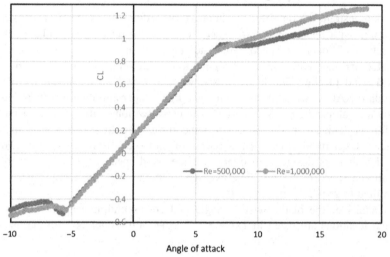

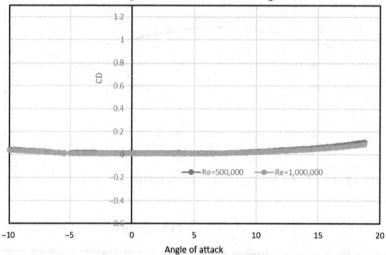

FIGURE 3.10 Variation of the lift and drag coefficients at various angle of attack of the selected airfoil (S809).

profile, experimental measurement of the lift and drag coefficients of the selected blade at various Reynolds number is also shown in Fig. 3.10 [2]. To evaluate the aerodynamic performance of the blade, one is required to carry out an analytical analysis using the BEM method. For the first step, the wind turbine is assumed to have three blades. The power loss at the tip and the drag forces are assumed to be negligible.

TABLE 3.3 Basic Information for the Input of BEM Analysis

Basic Information	
N (number of blades)	3.0000
R (radius)	5.0000
λ (tip speed ratio)	5.2360
U (wind speed)	7.2000
Ω Rotational speed (radius)	7.5398

First of all, based on the situation, one could obtain the basic design information for the blade. Note that there are some essential unit conversions within the steps. Since Microsoft© Excel spreadsheet uses radius as default, rotational speed is firstly expressed in radius. The basic information is tabulated in Table 3.3.

$$\Omega = \frac{2\pi \times 72}{60} = 7.5398$$

$$\lambda = \frac{\Omega R}{U_\infty} = 5.236 \tag{3.44}$$

Afterwards, we could determine the Reynolds number, solidity, tip speed ratio, and relative wind speed which are essential in the iterative process. Moreover, in calculating the Reynolds number, air properties are taken at $25°$ where the density and dynamic viscosity are 1.1839 kg/m^2 and 1.846×10^{-5} Pa s, respectively. Notice that the chord length is used as the characteristic length of the Reynolds number. Therefore, the Reynolds number is a variable along the span-wise of the blade. Calculation for the 5 m radial location is shown below. Calculations for other span wise location are presented in Table 3.4.

$$\lambda_r = \frac{\Omega r}{U_\infty} = 5.236$$

$$\sigma' = \frac{Nc}{2\pi r} = 0.0343$$

$$W = \sqrt{U_\infty^2 + \Omega^2 r^2} = 38.3805 \ m/s \tag{3.45}$$

$$\text{Re} = \frac{\rho U_\infty c}{\mu} = 883,706$$

With all the essential information, one could then start the iterative process with the initial guess. As discussed before, the guess value could be

TABLE 3.4 Aerodynamic Parameters at the Selected Span Wise Locations

Twisting and Tapered Blade	5 m	4 m	3.5 m	3 m	2.5 m
R (radius)	5.0000	4.0000	3.5000	3.0000	2.5000
c (Chord Length)	0.3590	0.4601	0.5096	0.5603	0.6110
γ (twisting angle)	−1.9746	−0.5247	0.4813	1.6470	2.2699
σ' (solidity)	0.0343	0.0549	0.0695	0.0892	0.1167
λ_r (local tip ratio)	5.2360	4.1888	3.6652	3.1416	2.6180
W (relative wind speed)	38.3805	31.0068	27.3540	23.7377	20.1779
Re (Reynolds number)	883,706	914,949	894,057	852,929	790,729

estimated from the following equations where it zero drag and zero tip loss are assumed.

$$\beta = \frac{2}{3}\tan^{-1}\left(\frac{1}{\lambda_r}\right) = 0.1258$$

$$a = \left(1 + \frac{4\sin^2\beta}{\sigma' C_L \cos\beta}\right)^{-1} = 0.3474 \qquad (3.46)$$

$$a' = \frac{1 - 3a}{4a - 1} = -0.1085$$

In above calculations, it is worth noting that the lift coefficient is determined based on the angle of the attack. Obviously, the angle of the attack i could be easily obtained by subtracting the relative wind direction β with the twisting angle γ. In principle, the lift coefficient is then evaluated by looking up table with simple interpolation. There are a number of ways to program the table look up in spreadsheet. One of the simplest way is to curve fitted the wind tunnel data with polynomials. Nonetheless, it is recommended not using a high degree polynomial which tends to be over-fitted the wind tunnel data.

As depicted in Table 3.5, the lift coefficient appears to have a mostly linear relationship with the angle of attack between $-5°$ to roughly $7°$. Utilize the trend line function in spreadsheet, one could easily obtain a line curve fitted function. For the Reynolds number of 500,000, the liner function between $-5°$ to $7°$ is given as:

$$y = 0.1157x + 0.1508 \qquad (3.47)$$

For the angle of attack beyond $7°$, a third-degree polynomial could be used as:

$$y = -0.000432x^3 + 0.016852x^2 - 0.190006x + 1.605458 \qquad (3.48)$$

Similarly, for the Reynolds number of 1,000,000, the liner function between $-5°$ and $6°$ is given as:

$$y = 0.1157x + 0.1508 \qquad (3.49)$$

For the angle of attack beyond $6°$, a third-degree polynomial could be used as:

$$y = -0.000142x^3 + 0.004322x^2 - 0.005378x + 0.783498 \qquad (3.50)$$

The lift coefficient could then be evaluated based on the above function with the IF function and simple interpolation according to the local Reynolds number. Notice that although only a third-degree polynomial was used, it is necessary to keep at least SIX decimal places for each coefficient to ensure the accuracy of the function. Here, some the guess a' values become negative

TABLE 3.5 Initial Guess Values With Respect to Each Span Wise Location

Initial Guess	5 m	4 m	3.5 m	3 m	2.5 m	2 m
β (°)	7.2083	8.9514	10.1740	11.7712	13.9370	17.0152
β (rad)	0.1258	0.1562	0.1776	0.2054	0.2432	0.2970
Angel of attack i	9.1829	9.4760	9.6926	10.1242	11.6671	13.8570
Lift coefficient (C_l)	0.9757	0.9889	0.9933	1.0009	1.0376	1.0829
$4\sin^2\beta$	0.0630	0.0968	0.1248	0.1665	0.2320	0.3425
$\sigma' C_L \cos\beta$	0.0332	0.0536	0.0680	0.0874	0.1175	0.1634
a	0.3451	0.3565	0.3526	0.3442	0.3362	0.3230
a'	−0.0928	−0.1631	−0.1408	−0.0865	−0.0250	0.1060

which in principle is invalid (i.e., induction factor ranges between 0 and 1). This error could be ignored at this stage as it is only an initial value.

Based on the initial guessed value, the iterative process could start for obtaining the solution. The process firstly starts from calculating the new relative wind direction β from the guessed value (Table 3.6):

$$\beta = \tan^{-1} \frac{(1-a)}{\lambda_r(1+a')} = 0.1370 \tag{3.51}$$

For the ease of calculating the angle of attack, the wind direction is converted in degree and the angle of attack is then evaluated. The corresponding lift coefficient is thereby updated. The new β is also used to determine the new axial and angular induction factors:

$$a = \frac{1}{(1 + 4\sin^2 \beta / \sigma' C_L \cos \beta)} = 0.3119 \tag{3.52}$$

$$a' = \frac{\sigma' C_L}{4\lambda_r \sin \beta}(1-a) = 0.0082 \tag{3.53}$$

The update values after first iteration for all span wise locations are listed in Table 3.7. For each iteration, it is crucial to monitor the progress of the iteration. One direct way is to monitor the change in iterative variation such Δa and $\Delta a'$ in this case. Comparing the new and old value of these variables, it could determine the progress of the iteration. In case all variables remain unchanged between iteration steps, the iteration is considered as converged.

The above iterative steps could be easily implemented in spreadsheet and repeated until a converged solution is obtained. The updated values for the 2nd iteration and the 10th iteration are tabulated in Tables 3.7 and 3.8, respectively.

As shown in the tables, after 10 iterations, the values of the axial and induction factors for all span wise locations remain unchanged and converged. This means the iterative process has obtained a converged solution for the given scenario. This also exemplifies the advantage of the iterative process. With the procedures, a complex system of equations could be easily resolved without going through complicated analytical derivation. The final converged solution is summarized in Tables 3.9 and 3.10.

With the converged solution is now possible to evaluate the performance of the whole blade. As discussed before, the power coefficient of the whole blade could be calculated as follow:

$$Cp = \frac{8}{\lambda^2} \int_{\lambda_h}^{\lambda} Q\lambda_r^3 a'(1-a)\left[1 - \frac{C_D}{C_L}\tan \beta\right]d\lambda_r \tag{3.54}$$

There are several ways for solving the integral. One of the simplest way is using trapezoidal rule to approximate the integral. In short, the trapezoidal

TABLE 3.6 Updated Aerodynamic Values After First Iteration

Iteration # 1	5 m	4 m	3.5 m	3 m	2.5 m	2 m
$\lambda_r(1+a')$	4.7501	3.5054	3.1493	2.8699	2.5526	2.3163
$(1-a)$	0.6549	0.6435	0.6474	0.6558	0.6638	0.6770
β (rad)	0.1370	0.1816	0.2028	0.2247	0.2544	0.2843
β (°)	7.8500	10.4021	11.6167	12.8718	14.5767	16.2919
i	9.8246	10.9268	11.1354	11.2249	12.3068	13.1336
C_L	0.9961	1.0389	1.0420	1.0366	1.0578	1.0625
New β (rad)	0.1370	0.1816	0.2028	0.2247	0.2544	0.2843
$\sigma'C_L \cos\beta$	0.0338	0.0561	0.0710	0.0901	0.1195	0.1610
$4\sin^2\beta$	0.0746	0.1304	0.1622	0.1985	0.2534	0.3148
$(4\sin^2\beta)/(\sigma'C_L \cos\beta)$	0.4534	0.4304	0.4375	0.4539	0.4715	0.5113
a	0.3119	0.3009	0.3044	0.3122	0.3204	0.3383
$\sigma'C_L$	0.0341	0.0571	0.0724	0.0924	0.1234	0.1677
$4\lambda_r \sin\beta$	2.8605	3.0252	2.9522	2.7994	2.6355	2.3502
a'	0.0082	0.0132	0.0171	0.0227	0.0318	0.0472
Δa	−0.0332	−0.0556	−0.0482	−0.0320	−0.0158	0.0153
$\Delta a'$	0.1010	0.1763	0.1578	0.1092	0.0568	−0.0587
New a	0.3119	0.3009	0.3044	0.3122	0.3204	0.3383
New a'	0.0082	0.0132	0.0171	0.0227	0.0318	0.0472

TABLE 3.7 Updated Aerodynamic Values After Second Iteration

Iteration # 2	5 m	4 m	3.5 m	3 m	2.5 m	2 m
$\lambda_r(1+a')$	6.8693	5.4491	4.7807	4.1224	3.4569	2.8030
$(1-a)$	0.8990	0.8237	0.8422	0.8908	0.9432	1.0587
β (rad)	0.1301	0.1500	0.1744	0.2128	0.2664	0.3612
β (°)	7.4559	8.5956	9.9907	12.1933	15.2614	20.6926
i	9.4305	9.1203	9.5094	10.5463	12.9916	17.5343
C_L	0.9834	0.9771	0.9873	1.0144	1.0791	1.1490
New β (rad)	0.1301	0.1500	0.1744	0.2128	0.2664	0.3612
$\sigma' C_L \cos \beta$	0.0334	0.0531	0.0676	0.0884	0.1215	0.1696
$4 \sin^2 \beta$	0.0674	0.0894	0.1204	0.1784	0.2771	0.4994
$(4 \sin^2 \beta / \sigma' C_L \cos \beta)$	0.4964	0.5938	0.5615	0.4955	0.4384	0.3397
a	0.3317	0.3726	0.3596	0.3313	0.3048	0.2535
$\sigma' C_L$	0.0337	0.0537	0.0686	0.0905	0.1259	0.1813
$4\lambda_r \sin \beta$	2.7177	2.5042	2.5435	2.6541	2.7565	2.9602
a'	0.0083	0.0134	0.0173	0.0228	0.0318	0.0457
Δa	**0.2307**	**0.1963**	**0.2018**	**0.2221**	**0.2480**	**0.3123**
$\Delta a'$	**−0.3037**	**−0.2874**	**−0.2871**	**−0.2894**	**−0.2887**	**−0.2926**
New a	0.3317	0.3726	0.3596	0.3313	0.3048	0.2535
New a'	0.0083	0.0134	0.0173	0.0228	0.0318	0.0457

TABLE 3.8 Updated Aerodynamic Values After 10th Iteration

Iteration # 10	5 m	4 m	3.5 m	3 m	2.5 m	2 m
$\lambda_r (1 + a')$	6.7279	5.4496	4.7522	4.0418	3.3395	2.6171
$(1-a)$	1.0000	1.0000	1.0000	1.0000	1.0000	1.0000
β (rad)	0.1476	0.1815	0.2074	0.2425	0.2910	0.3650
β (°)	8.4543	10.3981	11.8833	13.8969	16.6703	20.9116
i	10.4289	10.9227	11.4019	12.2499	14.4004	17.7533
C_L	1.0160	1.0388	1.0512	1.0706	1.1201	1.1497
new β (rad)	0.1476	0.1815	0.2074	0.2425	0.2910	0.3650
$\sigma' C_L \cos \beta$	0.0345	0.0561	0.0715	0.0927	0.1252	0.1695
$4 \sin^2 \beta$	0.0865	0.1303	0.1696	0.2307	0.3292	0.5096
$(4 \sin^2 \beta / \sigma' C_L \cos \beta)$	0.3985	0.4306	0.4216	0.4016	0.3804	0.3326
a	0.2849	0.3010	0.2966	0.2865	0.2756	0.2496
$\sigma' C_L$	0.0348	0.0570	0.0731	0.0955	0.1307	0.1815
$4\lambda_r \sin \beta$	3.0792	3.0241	3.0189	3.0181	3.0040	2.9902
a'	0.0081	0.0132	0.0170	0.0226	0.0315	0.0455
Δa	**0.0000**	**0.0000**	**0.0000**	**0.0000**	**0.0000**	**0.0000**
$\Delta a'$	**0.0000**	**0.0000**	**0.0000**	**0.0000**	**0.0000**	**0.0000**
New a	0.2849	0.3010	0.2966	0.2865	0.2756	0.2496
New a'	0.0081	0.0132	0.0170	0.0226	0.0315	0.0455

TABLE 3.9 Final Converged Solution Using BEM Method

Results	5 m	4 m	3.5 m	3 m	2.5 m	2 m
c (m)	0.3590	0.4601	0.5096	0.5603	0.6110	0.6611
λ_r	−1.9746	−0.5247	0.4813	1.6470	2.2699	3.1583
a	0.2849	0.3009	0.2965	0.2865	0.2755	0.2496
a'	0.0081	0.0132	0.0170	0.0226	0.0315	0.0455
i (°)	10.4311	10.9248	11.4039	12.2519	14.4026	17.7557
β (°)	8.4565	10.4001	11.8853	13.8988	16.6724	20.9140

TABLE 3.10 Power Coefficient Estimation Using Trapezoidal Rule

R	λ_r	a	a'	$\lambda_r^3 a'(1-a)$	h/2	Integral
5 m	5.2360	0.2849	0.0081	0.8303	0.5236	0.7895
4 m	4.1888	0.3009	0.0132	0.6775	0.2618	0.3318
3.5 m	3.6652	0.2965	0.0170	0.5898	0.2618	0.2851
3 m	3.1416	0.2865	0.0226	0.4992	0.2618	0.2380
2.5 m	2.6180	0.2755	0.0315	0.4098	0.2618	0.1895
2 m	2.0944	0.2496	0.0455	0.3139		
					Area under curve	1.8338
					Power coefficient, C_p	0.5351

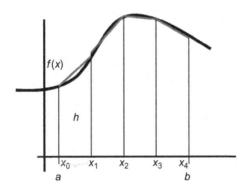

FIGURE 3.11   A schematic depiction of trapezoidal rule.

rule breaks the integral into a number of trapezoids and sum up the area of the each trapezoid to approximate the area under curve (Fig. 3.11).

With the converged solution is now possible to evaluate the performance of the whole blade. As discussed before, the power coefficient of the whole blade could be calculated as follow:

$$\int_a^b f(x)\, dx \approx \frac{b-a}{2n}\left(f(x_0) + 2f(x_1) + 2f(x_2) + \cdots + 2f(x_{n-1}) + f(x_n)\right) \quad (3.55)$$

For the power coefficient, for demonstration, only six span wise locations were shown in the calculation. Thererfore, using the trapezoidal rule, one could only utilize all six points as the function points and divide the area under curve for a total five sessions. Notice that the widths of each session are not uniform in this case. Moreover, the correction factor is equal to 1 assuming no loss at the tip. This also explain the slightly high power coefficient value.

REFERENCES

[1] Ingram G. Wind turbine blade analysis using the blade element momentum method. Note on the BEM method. Durham, NC, USA: Durham University; 2011.

[2] Somers, D.M. (1997) Design and experimental result for the S809 Airfoil. NREL/SR-440-6918, National Renewable Energy Laboratory.

[3] Hansen M. Aerodynamics of wind turbines. London, UK: Earthscan; 2008.

[4] NWTC Portal. NREL's S809 airfoil graphic and coordinates: https://wind.nrel.gov/airfoils/shapes/S809_Shape.html.

Chapter 4

Steady-State Heat Conduction Systems

4.1 APPLICATION OF HEAT TRANSFER PROCESS

In many engineering systems, energy transfer from one component to the other is a common encounter. One of the most common modes of energy transfer is the heat transfer process. In essence, regardless if you are an engineer or not, heat transfer is part of our daily life. Everyone would have the experience of waking up in a winter morning and rushing for a robe to cover your body. In the process, our body is exposed to cold air and ejecting heat to surroundings through heat transfer process. For minimizing the heat loss, additional clothing is utilized to reduce the heat transfer rate and preserve our energy.

The fundamental requirement for heat transfer is the existence of temperature difference between two media. Obviously, heat transfer could only occur from a high temperature medium (i.e., higher energy level) to a lower temperature medium. No heat transfer could occur if two media are at identical temperature. While heat transfer requirement seems to be simple and easily achieve, nevertheless, undesired heat transfer could inevitably occur in any engineering system as far as two media are at different temperature. Heat transfer therefore plays a significant role in many engineering designs, such as the radiator of car, electronic components of circuit board, heat exchanger of power plants, and many other real-life examples. Engineers or designers are in constant battle between controlling the heat transfer rate and performance optimization within limited budget. Determining and understanding the heat transfer is of paramount importance in almost every engineering systems.

4.2 THE THREE MODES OF HEAT TRANSFER

The prevalence of the heat transfer processes in engineering systems has been widely discussed in various engineering text books. The aim of this book has been focused on the numerical algorithm for solving engineering problem. To avoid distraction or confusion, a brief introduction to the basic equation for the modeling heat transfer is presented in this session. Readers interested in more fundamental in-depth knowledge of heat transfer process

Demystifying Numerical Models. DOI: https://doi.org/10.1016/B978-0-08-100975-8.00004-7

as well as its importance in thermodynamics analysis are diverted to other text books.

In a nutshell, heat transfer is the process where the form energy is exhibited in the form of heat and be transferred from one medium/system to another due to the exist of temperature difference. For most of the engineering applications, engineers are required to determine the rate of energy transfer for a proper system sizing. The term heat transfer frequently refers to the science in modeling the rates of energy be transferred from one to other-system in the form of heat. It is well known that heat transfer process can be categorized into three major modes: conduction, convection, and radiation.

4.2.1 Conduction

Conduction is a mode of heat transfer involving energy being transferred via interaction of particles of a medium or substance. Obviously, energy is transferred from the more energetic particles to a less energetic one. It is a common mistake for many people misunderstanding that conduction heat transfer is only occur in solid. From the theoretical viewpoint, as long as the particles exist, conduction could occur in any medium in any phase (e.g., solid, liquid, or gases). In solid phase, since all particles of substance are closely packed, energy is transferred through the combining effect of vibrations of molecules in a lattice structure and the motion of free electrons. In liquid or gases phase, energy is however transferred through collision or diffusion of molecules due to their random motion. Due to the difference of the energy transfer mechanism, one could notice that heat conduction would be more effective and dominant in solid phase. In liquid or gases phase, it will become more transparent in the following text that energy transfer would be dominant by convection indeed.

Considering conduction in a solid substance, as shown in Fig. 4.1, one could visualize that the rate of heat conduction depends on various factors such as thickness, surface area, and the material characteristic of the substance and the temperature difference driving the energy transfer. The rate of heat conduction could be expressed as follows:

$$\dot{Q} = \left[\frac{kA}{x}\right] [T_h - T_c] \tag{4.1}$$

where k is the thermal conductivity of the substance, x is the thickness of the medium, A is the cross-sectional area normal to the heat transfer direction, and T_h and T_c are the surface temperature at the hot and cold surfaces on the either side of the medium, as shown in Fig. 4.1.

The above equation forms a basic equation for evaluating the rate of heat conduction with simple geometrical and material information of the substance. This equation assumes that temperature difference between the hot and cold surfaces is uniform along the thickness direction. Removing such

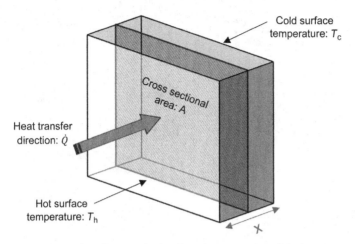

FIGURE 4.1 Visualization of heat conduction through a rectangular solid substance.

an assumption by expressing the temperature difference in terms of temperature gradient in x direction, the equation is then expressed as follow:

$$\dot{Q}_{conduction} = -kA\frac{dT}{dx} \tag{4.2}$$

which is commonly known as the Fourier's law of heat conduction. Since heat is conducted along the x direction when temperature is decreasing. One may notice that a negative sign is used to ensure the heat transfer rate to be positive along the x direction.

4.2.2 Convection

Convection is a physical phenomenon where energy is transferred from solid surface to the fluid either in liquid or gases phase. From the solid surface, there is two possible routes for energy transfer: conduction and fluid motion. The resultant energy transfer represents a combining effect of conduction and bulk fluid flow motion. Obviously, the motion of fluid transports mass and energy more effectively in compared with the conduction process. In most of the textbooks, the energy transported by the fluid motion is referred to convective heat transfer. In theoretical, the convective heat transfer is defined by the heat transfer across the surface in resulting from the motion of the fluid motion which is given by:

$$\rho A(C_v T)u_\infty \tag{4.3}$$

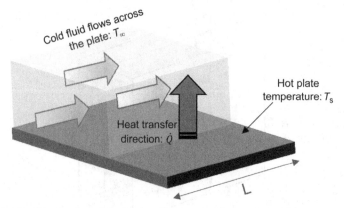

Cold fluid flows across the plate: T_∞

Hot plate temperature: T_s

Heat transfer direction: $\dot{Q}$

L

FIGURE 4.2 Visualization of convective heat transfer from a hot surface to a cold bulk fluid flow.

where ρ and C_v are the density and the specific heat capacity of the fluid, respectively, u_∞ is the free stream velocity of the fluid, and A is the area of the hot/cold surface, as shown in Fig. 4.2.

Combining the heat conduction, the overall heat transfer rate from the surface is then given by:

$$\dot{Q}_{convection} = -kA\frac{dT}{dx} + \rho A(C_v T)u_\infty \qquad (4.4)$$

One can notice from the equation that the effectiveness of the convective heat transfer large depends on the heat transfer due to the bulk fluid motion. The fluid flow characteristics have a significant effect on the resultant heat transfer rate. Thus the convection heat transfer can be classified into two different types according to the nature of the fluid flow. For the flow driven by external source (e.g., pump, fan, or wind), this type of heat transfer phenomenon is referred as forced convection. On the other hand, for the flow mainly driven by the buoyancy effect caused by the changes of density of fluid in the light of temperature rise or drop, such heat transfer behavior is referred to natural convection.

Since the fluid motion has a significant effect on the resultant heat transfer rate, to estimate the heat transfer properly, it requires information of how the fluid motion in the free stream be affected and create interaction with the surface. For a simple flat surface as shown in Fig. 4.2, a momentum and thermal boundary layer will be developed along the surface. Within the boundary layer, velocity and temperature distributions exhibit a highly nonlinear variation depending on various factors; including: local shear stress near the wall region, turbulence characteristics, buoyance effects, and fluid properties variations. For engineering application, the convection heat transfer rate is estimated by the following simplified equation:

$$\dot{Q}_{convection} = hA(T_\infty - T_s) \tag{4.5}$$

Here, the h represents the convection heat transfer coefficient which is normally estimated through empirical equations based on dimensionless analysis of a series of experimental data. T_∞ and T_s are the temperature of the free stream fluid and the temperature at the surface, respectively.

4.2.3 Radiation

Different with the previous two modes, heat transfer by the mean of radiation can occur without any medium. Radiation is transmitted by electromagnetic waves or photons caused by the changes of electronic configurations of an atom or molecules. Since electromagnetic waves travel at the speed of light, radiative heat transfer is the fastest mode of heat transfer. In fact, this is exactly how the energy be transferred from the sun to the earth.

It must be emphaized that most of the heat transfer studies only refer to thermal radiation, which is the electromagnetic wave emitted by any body, which is above absolute zero temperature. In other words, all bodies are practicallly emitted thermal radiation to surroundings; and simultenously absorbing or transmitting radiation from surroundings. For thermal radiation, the amount of radiation emitted by a body is proportional to its absolute temperature raised to the fourth power given as the Stefan-Boltzmann law:

$$\dot{Q}_{emit,radition} = \sigma A_s T^4 \tag{4.6}$$

where σ is the Stefan-Boltzmann constant and A_s is the surface area of a given body. However, depends on the surface property, the rate of emitted radiation could be different. For the ease of calculation, a surface property named as emissivity is introduced and related the rate of emitted radiation of every surface to an idealized blackbody. The blackbody is an idealized bady can emit radiation at the maximum rate. The emissivity of a blackbody surface is equal to 1. All other real surfaces emit radiation at a lower rate at the same given temperature which is expressed as:

$$\dot{Q}_{emit,radition} = \varepsilon \sigma A_s T^4 \tag{4.7}$$

The emissivity is a measure bounded between 0 and 1; representing a surface property and its relatively capacity of emitting thermal radiation in comparison to the idealized blackbody.

As mentioned earlier, radiative heat transfer in fact is a two-way interactive process between two or more sufaces of bodies. Radiation can be emitted from a surface of a body at any given angle across the whole solid angle (i.e., a sphere space). Meanwhile, the surface is simuultaneously absorbing

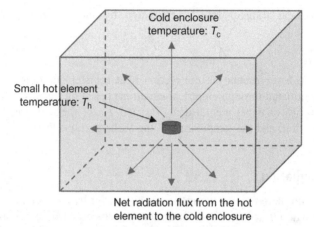

Net radiation flux from the hot
element to the cold enclosure

FIGURE 4.3 Visualization of the radiation emitted from a small hot body to the external cold enclosure.

thermal radiation from the other surfaces. According to the Kirchhoff's law of radation, the emissivity and absorptivity of a given surface at specific temperature and wavelenght is identical. Based on that, a simple relationship for the net radiation from a small hot body, of surface area A to a surrounding large cold enclosure, such as is illustrated in Fig. 4.3.

The radiative heat transfer is the net radiation from the hot surface to the cold enclosure. The net radiation occurs simply due to the fact that the cold surface emits less radiation at a lower temperature. Assuming that both hot and cold surfaces have the same emissivity ε, the radiative heat transfer is given by:

$$\dot{Q}_{radition} = \varepsilon \sigma A_s (T_h^4 - T_c^4) \qquad (4.8)$$

It is important to emphasize that the above equation embeds significant simplification in the formulation. In real situation, radiation is not uniformly emitted at all angles; same as the adsorption of radiation of any surface. Moreover, depends on the surface, the emitted radiation intensity could also dependent on the radiation wavelenght.

4.3 STEADY-STATE CONDUCTION PROBLEMS

A brief introduction to the three modes of heat transfer is presented in the previous section. Compared with the conduction heat transfer, convective and radiative heat transfer involve more complex physical phenomenon which makes the numerical modeling efforts becoming sophisticated. For example, convective heat transfer is significantly dependent on the near wall fluid motion and its associated turbulent flow (i.e., in most of the cases) structure within the boundary layer. In order to resolve the heat transfer rate,

one should firstly resolve the turbulent flow structure using advanced computational fluid dynamics technique. Similar argument also applies to the radiative heat transfer where radiation ray tracking and wavelength sensitive
emissivity, reflectivity, and transmissivity are usually required for practical
systems. Unfortunately, it requires lengthy discussion and complicated structured program which could not be covered in this book. The following content hereafter focuses only conduction heat transfer. Interested readers are
referred to other specific text books and the reference therein.

Conduction problems can be broadly categorized into two main types:
steady state and transient. For steady-state problems, the energy within the
system achieves a thermal equilibrium where energy input (i.e., heat source
or high temperature) is steadily conducted through the system and dissipated
away to surroundings with no energy accumulation or lost. In other words,
temperature at any point of the system remains constant and independent to
the time change. Steady-state conduction problems involve in many practical
engineering systems such as heat exchanger, insulation of pipe works, blast
furnace, printed circuit boards (i.e., PCBs) in electronics, power system, and
the like. In many of these applications, the complexity of the problem is
related to the complex geometry, materials, nonlinear thermodynamics properties, and heat transfer boundary conditions.

Previous research works have been carried out for developing analytical
methodology for obtaining solutions of conduction problem with given boundary
conditions and governing equations [1]. The analytical methods provide an accurate solution for some specific problem sets which could form a concrete foundation for verifying or validating the numerical methods. Unfortunately, analytical
solutions are very difficult to obtain for practical engineering systems. This leads
to implementation of numerical method to be discussed in this section.

4.3.1   Governing Equations

Following the discussion in the previous session, as shown in Fig. 4.1, heat
conduction rate through a simple one-dimensional material can be expressed
in simple differential equation. Assuming that the material mass and properties (e.g., thermal conductivity, density, and specific heat) are uniformly distributed (i.e., homogeneous), Fourier's law of heat conduction can be written
in differential form using the local temperature T as:

$$q_x = -k\frac{\partial T}{\partial x} \qquad (4.9)$$

To be more precise, the equation is now expressed in partial differential
form; considering heat conduction in x direction only. The q_x represents heat
transfer rate per unit area (i.e., heat flux) and time. Obviously, the heat flux
conducting through the media along the y or z direction (i.e., q_y and q_z) could
also be expressed in the similar partial differential form.

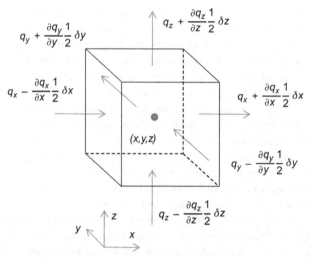

FIGURE 4.4 Energy additional caused by heat conduction through the surface of a rectangular element.

The governing equation for heat conduction is developed by applying the energy conservation law in a finite element. Considering a small rectangular element with the length of δx, the width of δy, and the depth of δz as shown in Fig. 4.4. According to the first law of thermodynamics, the rate of change of energy is equal to the sum of the rate of heat addition to the rectangular element. Notice that energy could also be added to the system in the form of work done due to fluid motion. Again, we only concern heat conduction in this section where energy change caused by the work done is not accounted as fluid motion is not involved in the element. The heat addition to the element can be caused by the conduction heat transfer across the surfaces of the element in the presence of temperature gradient (see also in Fig. 4.4).

As shown in the figure, the energy change due to thermal conduction in x-direction can be expressed as follow:

$$\left[\left(q_x - \frac{\partial q_x}{\partial x}\frac{1}{2}\delta x \right) - \left(q_x + \frac{\partial q_x}{\partial x}\frac{1}{2}\delta x \right) \right] \delta y\, \delta z = -\frac{\partial q_x}{\partial x}\delta x\, \delta y\, \delta z \qquad (4.10)$$

Furthermore, the energy change due to thermal conduction in y- and z-direction is expressed as:

$$\left[\left(q_y - \frac{\partial q_y}{\partial x}\frac{1}{2}\delta x \right) - \left(q_y + \frac{\partial q_y}{\partial x}\frac{1}{2}\delta x \right) \right] \delta y\, \delta z = -\frac{\partial q_y}{\partial x}\delta x\, \delta y\, \delta z \qquad (4.11)$$

$$\left[\left(q_z - \frac{\partial q_z}{\partial x}\frac{1}{2}\delta x \right) - \left(q_z + \frac{\partial q_z}{\partial x}\frac{1}{2}\delta x \right) \right] \delta y\, \delta z = -\frac{\partial q_z}{\partial x}\delta x\, \delta y\, \delta z \qquad (4.12)$$

In some cases, energy could also be added via volumetric heat source such as electrical resistance heating in PCBs. The additional volumetric heat is expressed as:

$$q_b V_{element} = q_b\, \delta x\, \delta y\, \delta z \qquad (4.13)$$

On the other hand, the rate of the increase of energy is given by the change rate of the internal energy (excluding kinetic energy as the element is stationary) as:

$$\frac{E_t - E_{t+\Delta t}}{\partial t} = \frac{\partial E}{\partial t} = \frac{\partial(mCT)}{\partial t} = \frac{\partial(\rho CT)}{\partial t} \delta x\, \delta y\, \delta z \qquad (4.14)$$

Noting that Eq. (4.14) represents the change of energy within the finite element with the time. Although this term becomes zero in steady-state condition, it is included at this stage for completeness in developing the governing equation.

Considering the energy conservation of the element, the additional energy input rate must be equal to the change rate of the energy contained in the element as given by:

$$\frac{\partial(\rho CT)}{\partial t} \delta x\, \delta y\, \delta z = - \left(\frac{\partial q_x}{\partial x} + \frac{\partial q_y}{\partial y} + \frac{\partial q_z}{\partial z} + q_b \right) \delta x\, \delta y\, \delta z \qquad (4.15)$$

Substitute the Fourier's law into the above equation and eliminate the volume of element from both sides gives:

$$\frac{\partial(\rho CT)}{\partial t} = \left[\frac{\partial}{\partial x}\left(k \frac{\partial T}{\partial x} \right) + \frac{\partial}{\partial y}\left(k \frac{\partial T}{\partial y} \right) + \frac{\partial}{\partial z}\left(k \frac{\partial T}{\partial z} \right) + q_b \right] \qquad (4.16)$$

This forms a generic heat conduction governing equation in a three-dimensional Cartesian coordinates space applicable for the transient conduction problem with internal volumetric heat source. For steady-state problems, the equation reduces to the following form:

$$\frac{\partial}{\partial x}\left(k \frac{\partial T}{\partial x} \right) + \frac{\partial}{\partial y}\left(k \frac{\partial T}{\partial y} \right) + \frac{\partial}{\partial z}\left(k \frac{\partial T}{\partial z} \right) + q_b = 0 \qquad (4.17)$$

Note that the governing equation could be reduced to one-dimensional or two-dimensional heat transfer by dropping the derivatives with respect to y or z direction from the equation.

4.4 BOUNDARY CONDITIONS FOR HEAT CONDUCTION PROBLEMS

The heat conduction equation developed above is obtained by applying energy conservation in a rectangular element. Although the governing equation is generic (i.e., applicable to any steady-state conduction problems), it contains absolutely no information regarding the condition of heat transfer problem that we need to solve. In other words, the governing equation is not well defined with information related to the practical surface condition such as surface temperature or heat flux. Mathematically, the governing equation consists of several partial differential terms. Solutions to these terms involve integration process that gives arbitrary constants leading to infinite number of solutions without additional information of the boundary condition. Without a proper boundary condition, it is impossible to obtain a unique solution that is relevant to the interested engineering problem. For example, considering the heat conduction through a rectangular block as shown in Fig. 4.1, temperature distribution within and the heat transfer rate through the block depends on the surface temperature at both the sides of the block. Imagine that the rectangular block represents the external wall separating outside and indoor environment. The surface temperature at both side depends on the outside weather and indoor temperature. Obviously, the heat transfer rate could substantially different changing from winter to summer. This clearly elucidates the importance of boundary condition and its involvement in defining an engineering problem.

4.4.1 Constant Temperature at Specific Surfaces

Following the above situation, for defining the heat transfer in winter or summer, it requires information of the weather condition and indoor environment. One of the most straight-forward way is measuring the surface temperature that exposes to the outside as well as the indoor conditions. Thus one of the boundary conditions commonly encounter in the practical engineering systems is specifying a constant temperature at the surface. Mathematically, such boundary condition is essential in any given problem. The specified temperature provides additional information for governing the arbitrary constant stem from integration process. The specified temperature could be implemented by simply specifying the temperature at the surface or even a particular point. For example, in summer time, the surface temperature measured at external wall is specified and expressed as a boundary condition as follow:

$$T_{x=0} = T_h = T_{measured} \qquad (4.18)$$

where $T_{x=0}$ represents the temperature at the location where the x coordinate equal to zero. Notice that the specified temperature could be specified uniformly over the whole surface or as any desired temperature distribution according to the practical system. Similarly, the specified temperature could remain constant for steady-state problem, while time-dependent temperature profile could also be specified for transient heat transfer discussion in the next chapter.

4.4.2   Adiabatic Condition at Well-Insulated Surfaces

In many practical systems, minimizing heat loss (or gain) from the system is a common consideration or measure for enhancing the system efficiency and performance. Therefore insulation is usually applied on a surface to thermally isolate the system from the surrounding influence. One should realize that there is no prefect insulation solution that could completely eliminate the heat transfer effect. Nonetheless, with the advancement of insultation material, it becomes a common practice in assuming a surface as adiabatic (i.e., zero heat transfer rate) as the surface is well insulated where the heat transfer could be suppressed to a negligible level. Different from specifying temperature, the adiabatic condition is expressed in terms of the temperature gradient or heat transfer rate via Fourier's law:

$$k\frac{\partial T}{\partial x}\bigg|_{x=0} = 0 \quad \text{or} \quad \frac{\partial T}{\partial x}\bigg|_{x=0} = 0 \qquad (4.19)$$

One could notice that the boundary condition is specified as the first-degree derivative. From the physical viewpoint, an adiabatic condition means that there is no heat transfer rate across the surface. Translating into mathematical expression implies either the thermal conductivity is zero or the temperature gradient is zero. It will become more transparent in later section how this boundary condition is implemented in numerical approach.

4.4.3   Surfaces Subjected to Convection or Radiation Heat Transfer

In many practical systems, it is common that some of the surfaces are exposed to the surrounding and subjected to convective or radiative heat transfer or both. For surface exposed to the air or surroundings, depends whether the surrounding air is externally driven or not, forced or natural convective heat transfer should be accounted. The boundary condition subjected to convective heat transfer is formulated based on energy balance. For simplicity, it is common to assume that the surface is infinite thin with negligible mass. In other words, there is no heat or energy stored in the surface. This implies that all energy passing through the surface must leave the system through the convective heat transfer. Putting the convective heat transfer in series with the heat conduction on the other side with the surface. Considering one-dimensional heat transfer in x-direction, if the wall surface is subjected to external convection at $x = 0$, the boundary condition is expressed as follow:

$$-k\frac{\partial T}{\partial x}\bigg|_{x=0} = h(T_\infty - T_{x=0}) \qquad (4.20)$$

Here, the h represents the convection heat transfer coefficient. Based on the equation, it is noting that the surface temperature at the boundary is now an unknown in the equation. Therefore, in order to proper define the problem, the temperature of the free stream fluid, T_∞, must be specified. Otherwise, the boundary is ill-defined where unique solution is not attainable. Moreover, it must be emphasized that the convective heat transfer coefficient also plays a significant in governing the temperature gradient caused by external convection. Not realistic specification of the convective heat transfer coefficient may cause stability or convergence problem in numerical methods.

Similarly, the surface could also expose to the radiative heat transfer when the radiation from surrounding is significant. Unfortunately, radiative heat transfer is mistakenly neglected in many cases. Many engineers believe that radiative heat transfer would be significant only at high temperature region (i.e., over 300°C). It is noteworthy that contribution of the radiative heat transfer could become significant at even very low temperature. For example, we all have the same experience feeling particularly cold in winter and feeling hot in summer while staying in indoor or a room that is maintained in same temperature throughout the year. The cold or hot feeling we experienced is caused by the radiative heat transfer between our body to the surroundings (especially outside surface through windows).

In fact, if a surface is subjected to natural convection as well as radiative heat transfer, radiative heat transfer could account significant portion in the overall heat transfer rate. Radiation could account around 30% of the total heat transfer when a person staying in an indoor environment that is not forced ventilated. Similar to the convection, for the wall is subjected to radiation $x = 0$, the boundary condition is expressed as follow:

$$-k\frac{\partial T}{\partial x}\bigg|_{x=0} = \varepsilon\sigma(T_{sur}^4 - T_{x=0}^4) \qquad (4.21)$$

Since the radiation has a fourth-order relationship to the temperature, the above boundary exhibits a nonlinear behavior in the problem. This could also impose additional difficulties in obtaining numerical solutions. In many practical systems, surface exposed to the surrounding is usually subjected to both convection and radiation heat transfer. A more generalized approach is to combine both influence in one boundary condition as follow:

$$-k\frac{\partial T}{\partial x}\bigg|_{x=0} = h(T_\infty - T_{x=0}) + \varepsilon\sigma(T_{sur}^4 - T_{x=0}^4) \qquad (4.22)$$

Obviously, additional considerations in the boundary condition would impose extra complexity in the calculation. This could become stability and convergence problem in the numerical procedures. More discussions regarding this will present in the following sections.

4.5 FINITE DIFFERENCE APPROACH

With the governing equations and boundary conditions, one could define a practical problem and translate it into mathematical formulation. If appropriate assumptions and engineering considerations were used, the formulation could accurate represent the realistic heat transfer behavior occurring in the practical systems. The only yet challenging task left behind is solving the governing equations for obtaining the solution.

Reviewing the governing equations revealed that the heat transfer problems are expressed mostly in terms of partial differential equations. There are many numerical approaches available for solving differential equations. Some of the widely adopted are finite difference, finite element, and finite volume methodology. It is important to point out that there is no the best approach after all or any particular approach is better than other. All approaches approximate the differential terms in algebraic equation that can be easily solving by properly structured program or even spreadsheet. In author's opinion, the choice of approach is a subjective matter rather than scientific consideration. In this chapter, we will adopt finite difference approach.

4.5.1 First-Order Finite Difference Approximation

Mathematically, a first derivative of a given function, (x), at a particular point can be interpreted as the slope of the tangent line to the curve of the function at the point. Fig. 4.5 shows the tangent line of a function curve

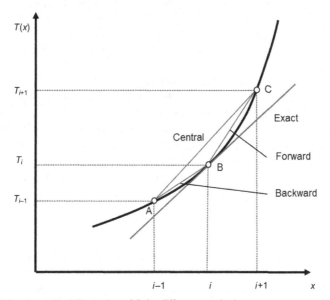

FIGURE 4.5 A graphical illustration of finite difference method.

(i.e., red color line) drawn at a particular point i. Here, the slope of the line represents the exact solution of the first derivative of the function. Obviously, it is impractical to find out the derivative in many cases. This is simply because the function of the given variable is unknown which we need to obtain. Back in our heat transfer problem, the derivative of temperature means the exact solution of the temperature gradient at a particular point where temperatures are the answer we need to find out at first place.

Without knowledge of the exact function, at best, one could only approximate the solution based on what information is available. Assuming that one could obtain the temperature value at points A, B, and C as shown in Fig. 4.5, the slope of the tangent line can be approximated by jointing points B and C. The derivative is then estimated as:

$$\frac{dT(x_i)}{dx} \cong \frac{T_{i+1} - T_i}{x_{i+1} - x_i} \tag{4.23}$$

With no doubt, the approximation above is far from prefect and contains error inevitably. The error associated to such approximation could be elucidated via Taylor series expansion. According to the theorem, the function value at point C could be obtained without any error with the series expansion:

$$T(x_{i+1}) = T(x_i) + \Delta x \frac{dT(x_i)}{dx} + \frac{1}{2!}\Delta x^2 \frac{d^2T(x_i)}{dx^2} + \frac{1}{3!}\Delta x^3 \frac{d^3T(x_i)}{dx^3} + \ldots \tag{4.24}$$

From the above equation, one could obtain the exact solution of first derivative by separating the second term on the right-hand side which is given as:

$$\frac{dT(x_i)}{dx} = \frac{T(x_{i+1}) - T(x_i)}{\Delta x} - \frac{1}{2!}\Delta x \frac{d^2T(x_i)}{dx^2} + \frac{1}{3!}\Delta x^2 \frac{d^3T(x_i)}{dx^3} + \ldots \tag{4.25}$$

Obviously, the approximation based on Eq. (4.23) is done by neglecting all the high-order term inside the red box. The first term of the omitted term is proportional to the Δx. In literature, the above approximate is referred as first-order approximation. This is named according to the order of the first omitted term. One the other hand, considering the coordinate sequence, as the approximation is done by joining the next point in x direction, it also refers as forward differencing scheme. Similarly, we could also approximate the first derivative by jointing points A and B which is given by:

$$\frac{dT(x_i)}{dx} \cong \frac{T_i - T_{i-1}}{x_i - x_{i-1}} \tag{4.26}$$

This is commonly referred as backward differencing scheme due to a point behind in x direction is used. Fig. 4.5 exemplified the approximation error associated with the forward and backward differencing scheme. It can be observed that the slope of the A-B line and B-C line is different from the exact solution. It must be emphasized that the figure is only an example for illustration. In theory, both forward and backward scheme are first-order approximation with the error proportional to the Δx. This explains why the prediction error of numerical method depends on the mesh size (i.e., spacing between discrete points such as A and B). One of the effective way to minimize the approximation error is reducing the mesh size (i.e., Δx). Unfortunately, this would induce more computational steps and time in calculation. With the ever-increasing computational power, it seems to be a reasonable choice since computational power is improving rapidly in the past decades.

4.5.2   Second-Order Finite Difference Approximation

Another possible way to improve the accuracy is employing a higher order differencing scheme. Instead of jointing the next or behind point, the approximation could be done jointing the points A and C. The derivative is estimated as:

$$\frac{dT(x_i)}{dx} \cong \frac{T_{i+1} - T_{i-1}}{x_{i+1} - x_{i-1}} = \frac{T_{i+1} - T_{i-1}}{2\Delta x} \tag{4.27}$$

As illustrated in Fig. 4.5, the resultant of A-C line appears to have a slope closer to the exact solution. This is benefit from the additional information drawing from both upstream and downstream points. Since both upstream and downstream points are employed, this approximation method is commonly referred as central scheme. Mathematically, the error of the central scheme can also be estimated based on the Taylor series expansion. For the backward scheme, the Taylor series is given by:

$$T(x_{i-1}) = T(x_i) - \Delta x \frac{dT(x_i)}{dx} + \frac{1}{2!}\Delta x^2 \frac{d^2T(x_i)}{dx^2} - \frac{1}{3!}\Delta x^3 \frac{d^3T(x_i)}{dx^3} + \dots \tag{4.28}$$

By subtracting Eq (4.28) and Eq. (4.29) gives the following expression:

$$\frac{dT(x_i)}{dx} = \frac{T(x_{i+1}) - T(x_i)}{2\Delta x} + \frac{1}{3!}\Delta x^2 \frac{d^3T(x_i)}{dx^3} + \dots \tag{4.29}$$

As shown in the above equation, the first term of the omitted errors (i.e., highlighted in red color) is proportional to the Δx^2. In other word, the central differencing scheme has a second-order accuracy. This also explains why the approximation could yield a more accurate result in comparison to the forward and backward schemes.

Nevertheless, heat conduction governing equations involve second-order derivatives. With reference to the differencing schemes of first-order derivative, we could approximate the second order by summing Eqs. (4.28) and (4.24) which give the expression as follow:

$$\frac{d^2T(x_i)}{dx^2} = \frac{T(x_{i+1}) - 2T(x_i) + T(x_{i-1})}{\Delta x^2} + \frac{1}{4!}\Delta x^2 \frac{d^4T(x_i)}{dx^4}\dots \quad (4.30)$$

Similar to the first derivative, the high-order terms in the Taylor's series are omitted for approximation:

$$\frac{d^2T(x_i)}{dx^2} \cong \frac{T(x_{i+1}) - 2T(x_i) + T(x_{i-1})}{\Delta x^2} \quad (4.31)$$

By using the above approximation, the second derivatives of temperature appeared in the governing equation can be replaced by simple algebraic equation expressed in modal points. The above procedure is usually referred as discretization. Although only one-dimensional derivatives are considered in this section. The above discretization can be easily extended to two- or three-dimensional space.

Moreover, it can also be seen in the formulation that the discretization is applicable to any node point; including node points adjacent to the boundary conditions. The reader is reminded that boundary condition is simply additional information to define the problem. The boundary condition should also subject to the governing equation which in turn subject to the discretization as well. Nevertheless, there are some considerations in regard the implementation of the boundary condition with discretized equation as discussed in the following example.

4.6 ONE-DIMENSIONAL STEADY-STATE CONDUCTION

With the discretization scheme and the governing equations, it is now possible to solve a heat conduction problem with the finite difference formulation. In this section, or simplicity, the implementation of finite difference numerical method will be presented in a one-dimensional steady-state problem. A brief description of the problem example is firstly discussed in the following.

4.6.1 One-Dimensional Conduction With Internal Heat Source

To demonstrate the method, we first consider steady one-dimensional heat conduction through a solid block of thickness L where heat is also uniformly generated internally. One side of the solid black is subjected to external heat input (i.e., either constant surface temperature or constant heat flux). The other side of the block is losing energy to the surroundings where surface is kept at a constant low temperature.

With the one-dimensional conduction, the governing equation can be expressed by dropping the other terms in y and z directions from Eq. (4.17) as follow:

$$\frac{\partial}{\partial x}\left(k\frac{\partial T}{\partial x}\right) + q_b = 0 \qquad (4.32)$$

To discretize the above governing equation, the solid block is divided into a number of discrete point (also refers as mesh in other text book). As shown in Fig. 4.6, a total of five nodal points is evenly distributed along the x direction within the block where the distance between point equals to Δx. The nodes 1 and 5 are connected to the boundary (i.e., boundary condition) which is an imaginary node outside the block with the same mesh spacing. It must be to emphasize that some text book treats the boundary condition by placing a nodal point exactly on the surface. More discussions will follow in the next section.

First of all, we focus on developing a generic formulation for all internal nodal points (i.e., points 2−4). Considering the nodal point 3, the surfaces of the control volume are placed at the middle between two nodal points. Heat is being conducted into the control volume through on the surface on the left and right. Meanwhile, heat is also generated within the control volume (Fig. 4.7).

One should notice that the physical interpretation of the governing Eq. (4.32) refers to the energy balance of any given control volume. The second derivative of temperature models the heat conduction into the control volume through the left and right surfaces. The governing equation is then approximated as:

$$\left(k\frac{\partial T}{\partial x}\right)_{right} - \left(k\frac{\partial T}{\partial x}\right)_{left} + q_b = 0 \qquad (4.33)$$

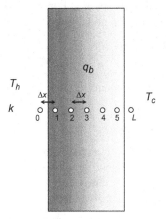

FIGURE 4.6   Modeling of a block losing energy.

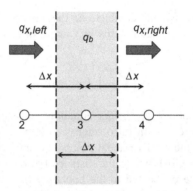

FIGURE 4.7 Heat balance considerations for the internal nodal points.

Following the first-order differencing scheme discussed above, the temperature gradient at the left and right is discretized as follow:

$$kA\frac{T_4 - T_3}{\Delta x} - kA\frac{T_3 - T_2}{\Delta x} + q_b\Delta xA = 0 \tag{4.34}$$

Note that heat conductions and generation are expressed in terms of heat flux and volumetric heat source in the governing equation, respectively. The equation is multiplied by the volume of the internal nodal point. Therefore, surface area (i.e., A) and mesh distance Δx appear in the above equation. As one-dimensional problem is considering, we could simply assume the surface area as unity. Expanding the first and second terms of the equation gives:

$$kA\left(\frac{T_4 - 2T_3 + T_2}{\Delta x}\right) + q_b\Delta xA = = 0 \tag{4.35}$$

The governing equation is now expressed as algebraic equation without any derivative and differential equation. Notice that the discretized equation is applicable to all internal nodal points. For more generic formulation, one could express the equation for the nodal point i as:

$$kA\left(\frac{T_{i+1} - 2T_i + T_{i-1}}{\Delta x}\right) + q_b\Delta xA = = 0 \tag{4.36}$$

For the ease of solving the equation later, it is common to rearrange the equation in form as following:

$$a_iT_i = a_{i+1}T_{i+1} + a_{i-1}T_{i-1} + S \tag{4.37}$$

where a_i, a_{i-1}, and S are the coefficients of the resultant simultaneous equation. Notice that such rearrangement aims to express the simultaneous equations of all nodal points in terms of matrix expression. With the rearrangement, the coefficients are given as:

$$a_{i+1} = a_{i-1} = \frac{kA}{\Delta x}, \quad a_i = a_{i-1} + a_{i+1} = \frac{2kA}{\Delta x} \quad \text{and} \quad S = q\Delta xA \tag{4.38}$$

From the above equation, it can be observed that the coefficient relates to the value of thermal properties, mesh spacing, and surface area of the control volume. This information is readily available once a computational mesh is generated for a practice problem. The only unknown in the equation is temperature value at the nodal points. Again, these formulations are insufficient for obtaining a unique solution with proper boundary conditions as introduced in the following.

4.6.2   Formulation for the Boundary Conditions

In section 4.4, we have introduced some common encountered boundary conditions in practical engineering systems. Nonetheless, the formulation of the boundary condition depends on the arrangements and considerations of the discretization. This section devotes to introduce the concept and technique which is commonly adopted in the numerical method.

From the mathematical viewpoint, as the governing equation consists of second-order derivatives, boundary condition can be specified as the zero- and first-order derivatives. The zero- and first-order derivatives in engineering terms are the temperature and heat flux, respectively. As mentioned earlier, attention should be paid to the mesh or nodal point arrangement at the boundary condition. In general, two common approaches are widely adopted. One is generating a uniform mesh space even at the boundary which is allocated as an imaginary node. This approach would provide convenience in handling mesh spacing and specifying heat flux at the boundary. Unfortunately, it will become more complicated in specifying a constant temperature boundary.

Another approach is to generate the mesh node exactly on the surface, as shown in Fig. 4.8. The advantage of this approach is the desired boundary condition could be specified at the exact location of the surface. This is

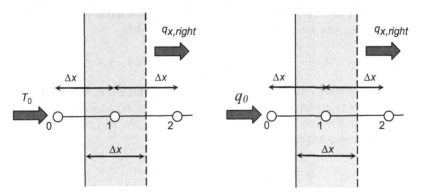

FIGURE 4.8   Mesh arrangement at the constant temperature and heat flux boundary condition.

particularly convenient when constant temperature is specified at the surface. Nevertheless, the mesh spacing reduces to half; leading to a slightly different formulation in the discretized equation.

The **specified temperature** boundary condition is the simplest type for implementation. As shown in Fig. 4.8, the constant temperature can be specified at the nodal point 0 at the surface. Considering the energy balance for the node 1, the discretized governing equation becomes as follow:

$$kA\frac{T_2 - T_1}{\Delta x} - kA\frac{T_1 - T_0}{\Delta x/2} + q_b \Delta x A = 0 \tag{4.39}$$

For rearranging the equation, since T_0 is a given value as boundary condition, we set the coefficient $a_0 = 0$ and associate the corresponding terms with the coefficient S. Details of the formulation are shown as follow:

$$a_2 = \frac{kA}{\Delta x}, \quad a_0 = 0, \quad a_1 = \frac{kA}{\Delta x} + \frac{2kA}{\Delta x} = \frac{3kA}{\Delta x} \quad \text{and} \quad S = q_b \Delta x A + \frac{2kA}{\Delta x} T_0 \tag{4.40}$$

Similarly, in case temperature is specified at the other end (i.e., $x = L$), the discretized governing equation for node 5 becomes as follow:

$$kA\frac{T_L - T_5}{\Delta x/2} - kA\frac{T_5 - T_4}{\Delta x} + q_b \Delta x A = 0 \tag{4.41}$$

After rearranging the equation, the coefficient $a_L = 0$ with the corresponding terms associated with the coefficient S. The formulation becomes as follow:

$$a_L = 0, \quad a_4 = \frac{kA}{\Delta x}, \quad a_5 = \frac{kA}{\Delta x} + \frac{2kA}{\Delta x} = \frac{3kA}{\Delta x} \quad \text{and} \quad S = q\Delta x A + \frac{2kA}{\Delta x} T_L \tag{4.42}$$

In case **constant heat flux** is specified at the boundary as shown in Fig. 4.8. For node 1, the formulation is quite similar to the internal nodal point, except the temperature at the boundary is now unknown with a half mesh spacing. The formulation is therefore expressed as:

$$a_2 = \frac{kA}{\Delta x}, \quad a_0 = \frac{2kA}{\Delta x}, \quad a_1 = a_0 + a_2 = \frac{3kA}{\Delta x} \quad \text{and} \quad S = q\Delta x A \tag{4.43}$$

As temperature at the surface is unknown, it is necessary to add another simultaneous equation to obtain the solution. Therefore, additional equation for node 0 is now required. Considering the node 0, the energy conservation can be expressed as follow:

$$kA\frac{T_1 - T_0}{\Delta x/2} - q_0 A + q_b \frac{\Delta x}{2} A = 0 \tag{4.44}$$

It is also straight-forward to rearrange the equation in terms of coefficient as follow:

$$a_1 = a_0 = \frac{2kA}{\Delta x} \quad and \quad S = q\frac{\Delta x}{2}A + q_0A \tag{4.45}$$

For adiabatic boundary with insulation, the above formulation could then further reduce by setting the heat flux $q_0 = 0$. Furthermore, it is also possible to specific a **convective heat flux** or **radiative heat flux** at the surface. To be consistent with formulation, we define heat flux transferring along the x direction as negative. The convective and radiation heat transfer is given as:

$$q_0 = h(T_0 - T_\infty) \quad or \quad q_0 = \varepsilon\sigma(T_0^4 - T_\infty^4) \tag{4.46}$$

Considering the boundary is subject to convective heat flux, the energy conservation is given by:

$$kA\frac{T_1 - T_0}{\Delta x/2} - hA(T_0 - T_\infty) + q_b\frac{\Delta x}{2}A = 0 \tag{4.47}$$

where the rearranged coefficient is given as:

$$a_1 = \frac{2kA}{\Delta x}, \quad a_0 = \frac{2kA}{\Delta x} + hA \quad and \quad S = q\frac{\Delta x}{2}A + hAT_\infty \tag{4.48}$$

One should notice that the surrounding temperature should be clearly defined in the formulation. Otherwise, the boundary condition is ill-defined where it is not possible for obtaining the unique solution.

4.6.3   Heat Dissipation of a Hot Plate

To illustrate the solution method, we first consider steady one-dimensional heat conduction through a hot plate with the thickness of 5 cm. The hot plate has an internal electric heating element which generates heat uniformly within the hot plate at a constant rate of 3×10^6 W/m^3. The thermal conductivity of the hot plate is a constant of $k = 28$ W/m. One side of the hot plate is attached to the base which is well insulated with negligible heat loss. A container is putting on the other side with a large block of ice at the temperature of 0°C inside. Estimate the temperature distribution with the hot plate under steady-state condition (Fig. 4.9).

The problem can be solved by using finite difference method with a total of five nodal points insider the hot plate. The mesh spacing between nodal points is uniform of 1 cm where half mesh spacing of 0.5 cm is adopted at the boundary nodes. Assuming one-dimensional heat transfer, the surface area of all control volume is considered as unity. Following

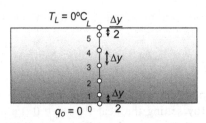

FIGURE 4.9 A schematic for the heat dissipation of a hot plate and the distribution of the computational node points.

Eq. (4.38), the coefficients for all internal points (e.g., nodes 1–4) are given by:

$$a_{i+1} = a_{i-1} = \frac{kA}{\Delta x}, \quad a_i = a_{i-1} + a_{i+1} = \frac{2kA}{\Delta x} \quad \text{and} \quad S = q\Delta xA$$

$$a_{i+1} = a_{i-1} = \frac{(28)(1)}{0.01} = 2800, \quad a_i = 5600 \quad \text{and} \tag{4.49}$$

$$S = 3 \times 10^6 \times 0.01 \times 1 = 30,000$$

For the node 5 which is connected to the surface maintained at 0°C, the coefficients are expressed as:

$$a_L = 0, \quad a_4 = \frac{kA}{\Delta x}, \quad a_5 = \frac{kA}{\Delta x} + \frac{2kA}{\Delta x} = \frac{3kA}{\Delta x} \quad \text{and} \quad S = q\Delta xA + \frac{2kA}{\Delta x}T_L$$

$$a_L = 0, \quad a_4 = 2800, \quad a_5 = 8400 \quad \text{and} \quad S = 30,000 + 5600 \times 0 = 35,600 \tag{4.50}$$

For the node 1, considering only half mesh spacing, the coefficients are expressed as:

$$a_2 = \frac{kA}{\Delta x}, \quad a_0 = \frac{2kA}{\Delta x}, \quad a_1 = a_0 + a_2 = \frac{3kA}{\Delta x} \quad \text{and} \quad S = q\Delta xA \tag{4.51}$$

$$a_2 = 2800, \quad a_0 = 5600, \quad a_1 = 8400 \quad \text{and} \quad S = 30,000$$

For the node 0 which is well insulated as adiabatic condition, the coefficients are expressed as:

$$a_1 = a_0 = \frac{2kA}{\Delta x} \quad \text{and} \quad S = q\frac{\Delta x}{2}A + q_0A \tag{4.52}$$

$$a_1 = a_0 = 5600 \quad \text{and} \quad S = 15,000 + 0 \times 1 = 15,000$$

The resultant simultaneous equations for all nodal points can be expressed as follow:

$$5600T_0 = 5600T_1 + 15,000$$
$$8400T_1 = 2800T_2 + 5600T_0 + 30,000$$
$$5600T_2 = 2800T_3 + 2800T_1 + 30,000$$
$$5600T_3 = 2800T_4 + 2800T_2 + 30,000$$
$$5600T_4 = 2800T_5 + 2800T_3 + 30,000$$
$$8400T_5 = 2800T_4 + 35,600$$

The above set of equation could then be expressed in terms of matrix formulation as follow:

$$[A]\ [T] = [S] \tag{4.53}$$

With substitution, the matrix become

$$
\begin{bmatrix}
5600 & -5600 & 0 & 0 & 0 & 0 \\
-5600 & 8400 & -2800 & 0 & 0 & 0 \\
0 & -2800 & 5600 & -2800 & 0 & 0 \\
0 & 0 & -2800 & 5600 & -2800 & 0 \\
0 & 0 & 0 & -2800 & 5600 & -2800 \\
0 & 0 & 0 & 0 & -2800 & 8400
\end{bmatrix}
\begin{bmatrix}
T_0 \\ T_1 \\ T_2 \\ T_3 \\ T_4 \\ T_5
\end{bmatrix}
=
\begin{bmatrix}
15,000 \\ 30,000 \\ 30,000 \\ 30,000 \\ 30,000 \\ 35,600
\end{bmatrix}
$$

The temperature of all nodal points can now be obtained by matrix operations. The matrix operation involves the calculation of matrix inverse as follow:

$$[A]^{-1}[A]\ [T] = [A]^{-1}[S] \tag{4.54}$$

Although it is tedious to calculate the matrix inverse, most of spreadsheet software are now equipped with a the built-in function for such operation. For Microsoft © Excel users, the matrix inverse could be obtained by following steps:

- Key in the coefficient matrix [A]. For our case, it is a 6×6 matrix in the cells ranging from B2:G7
- Highlight a range of cells where the 6×6 matrix inverse should be located (e.g., H2:M7)
- Key in the command = MINVERSE(B2:G7) in function window
- Press the Ctrl and the Shift keys at the same time and press Enter.

The matrix inverse is then obtained as presented in Table 4.1.

The final step for obtaining the solution is multiplying the matrix inverse with the coefficient vector [S]:

$$[T] = [A]^{-1}[S] \tag{4.55}$$

TABLE 4.1 The Coefficient of the Matrix Inverse

Matrix Inverse

0.001786	0.001607	0.00125	0.000893	0.000535714	0.000178571
0.001607	0.001607	0.00125	0.000893	0.000535714	0.000178571
0.00125	0.00125	0.00125	0.000893	0.000535714	0.000178571
0.000893	0.000893	0.000893	0.000893	0.000535714	0.000178571
0.000536	0.000536	0.000536	0.000536	0.000535714	0.000178571
0.000179	0.000179	0.000179	0.000179	0.000178571	0.000178571

TABLE 4.2 The Final Temperature Distribution for the Given Heat Conduction Example

Final Answer

T_0	161.7142857
T_1	159.0357143
T_2	142.9642857
T_3	116.1785714
T_4	78.67857143
T_5	30.46428571

For Microsoft © Excel users, the matrix inverse could be obtained by following steps:

- Highlight a range of cells where the temperature vector $[T]$ should be located (e.g., O2:O7)
- Key in the command = MMULT(H2:M7, O2:O7) in function window
- Press the Ctrl and the Shift keys at the same time and press Enter.

The final temperature distribution within the hot plate can be obtained in Table 4.2.

In this section, the finite difference formulation for a one-dimensional heat conduction problem is introduced. The finite difference approach approximates the derivatives in the heat transfer equation with algebraic equation. As shown in the above example, the algebraic equation forms a set of simultaneous equations that can be solved by matrix operations using spreadsheet.

The matrix size depends on the number of mesh node. For our case, even for a one-dimensional problem, six mesh nodes result in a 6×6 matrix which is manageable for matrix operation. Nonetheless, in case the problem is extended to two- or even three-dimensional heat transfer. This is not viable to manage such a large matrix. In fact, the solution of larger system can be obtained based on iterative processes. Details of the iterative procedure together with the method for solving time-dependent heat transfer problem is discussed in the next chapter.

REFERENCE

[1]   Taler, J., Duda, P. (2006) Solving direct and inverse heat conduction problems; Chapter 6, Pages 141–160; ISBN 978-3-540-33471-2.

Chapter 5

Two-Dimensional and Transient Heat Conduction

5.1 IMPORTANCE OF TWO-DIMENSIONAL AND TRANSIENT HEAT CONDUCTION PROBLEMS

In the previous chapter, finite difference method for solving the one-dimensional steady state heat conduction systems has been presented. Although it only focuses on the heat conduction systems, boundary conditions coupling the convective and radiative heat transfer are also discussed.

Nonetheless, in many engineering systems, heat conduction exhibits predominantly in two- or three-dimensional nature. Engineering systems consist of multiple components where thermal properties could be significantly different, leading to second- or third-dimension temperature gradient and its corresponding heat transfer. In some cases, inhomogeneous temperature distributions could lead to substantial or even significant thermal stress to the systems—causing deformation, crack, or even fracture. A simple daily example of such phenomenon is the cracking of a glass cup when suddenly heat is imposed inside. When hot water pouring inside the cup, the inner glass surfaces are subjected to high temperature. However, due to the low thermal conductivity, outer glass surfaces are still maintained in low temperature. The unevenly temperature distribution causes significant thermal stress that eventually cracks the glass cup.

Solving the two-dimensional heat conduction could obtain useful temperature distribution for engineers and design to optimize their designs. Since most of the engineering systems are designed for long-term operation, design considerations are commonly focused on the steady state temperature distribution. Nevertheless, in some situation (especially for high-power devices), transient heat generation and conduction could incur significant temporal thermal load. It is also important to resolve the transient heat conduction to obtain the time-dependent temperature generation and dissipation characteristics.

Demystifying Numerical Models. DOI: https://doi.org/10.1016/B978-0-08-100975-8.00005-9

5.2 DIRECT VERSUS ITERATIVE METHODS

Through the finite difference method, heat conduction governing equation can be discretized into a set of simultaneous algebraic equations. Solutions of the simultaneous equations can be obtained by simple matrix inversion function available in spreadsheets. Such matrix inversion procedures are referred as direct method in literature. As demonstrated in Chapter 4, the matrix inversion procedures require user manually formulate and input the coefficient matrices. Referring the problem example in Chapter 4, a one-dimensional problem with six mesh nodes results in a 6×6 coefficient matrix. For simple one-dimensional steady state heat conduction, the solution procedures seem to be manageable for most applications. Nevertheless, for two-dimensional problems, it is common to have 10 or even 100 nodes in one domain. The large number of mesh results a significant large coefficient matrix, marking the formulation and manual input procedures unbearable (i.e., 10 nodes require a 10×10 matrix with 100 coefficients).

To eliminate the tedious procedures in direct method, for large number of mesh, it is common to adopted iterative method where the solutions are obtained by a "trial-and-error" procedure with an initial-guess. In theory the direct and iterative methods should yield identical solutions. Obviously, this assumes that a converged solution could be obtained through the iterative method. To ensure an accurate result could be obtained, one should specify a convergence target to terminate the iterative procedures. With the advancement of software, aforementioned iterative procedures could also be implemented easily in the spreadsheet. Implementation of such two-dimensional heat conduction iterative method is discussed in this chapter.

5.3 TWO-DIMENSIONAL STEADY STATE HEAT CONDUCTION

In this session, following the one-dimension heat conduction problem, we will consider extending our formulation to two-dimensional steady state conduction problem in rectangular domain. Finite difference method is adopted to discretize the heat conduction governing equation. Demonstrating the formulation aims in twofold, readers can follow similar formulation procedures extending the problem to three-dimensional cases; it presents the mathematical foundation for implementing the iterative procedures.

5.3.1 Two-Dimensional Discretization

We firstly consider the steady two-dimensional heat conduction through a rectangular solid block where temperature variation is significant in both x- and y-directions. Aiming to determine the temperature distribution in the two-dimensional space, a numerical mesh is needed dividing the rectangular

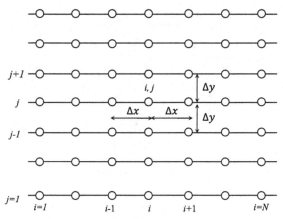

FIGURE 5.1   A two-dimensional numerical mesh for the finite difference discretization.

domain in a number of nodal points with the spacings of Δx and Δy along the x and y direction. A visualization of the numerical mesh is depicted in Fig. 5.1.

Since the formulation for all internal nodes is identical, it is convenient to refer the nodal point based on numbers or notations instead of actual physical location. As depicted in Fig. 5.1, it is common to use double subscript notation (i, j) to number a two-dimensional numerical mesh. In x-direction, nodal point is notated as $i = 1, 2, 3 \ldots, N$, while $j = 1, 2, 3 \ldots, M$ is counted along the y-direction. For a uniformly spaced mesh the actual coordinates of the nodal point can be easily calculated based on the nodal number. One should notice that choosing double subscript for notation is mainly for the convenience of programming. Most of spreadsheet calculations are based on cells that are also notated in double subscription using columns (i.e., A, B, ..., Z and more) and rows (i.e., 1, 2, 3, ..., 10 and more). The following formulation could be also easily referred to cells operation working in spreadsheet.

Similar to the one-dimensional problem, we focus on developing a generic formulation for all internal nodal points (i.e., point (i, j)). Considering the nodal point (i, j) with the volume of $\Delta x \times \Delta y \times 1$, the surfaces of the control volume are placed at the middle among all nodal points. Heat is being conducted into the control volume through the surface from both vertical and horizontal direction. Meanwhile, heat is also generated within the control volume (Fig. 5.2).

Considering the heat balance on the control volume, the governing equation can be expressed as

$$q_{x,\text{left}} - q_{x,\text{right}} + q_{y,\text{bottom}} - q_{y,\text{top}} + q_b = 0 \qquad (5.1)$$

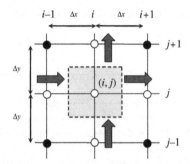

FIGURE 5.2   Heat balance considerations for the two-dimensional internal nodal points.

Based on the Fourier's law of heat conduction, the heat balance is given by

$$\left(kA \frac{\partial T}{\partial x} \right)_{\text{left}} - \left(kA \frac{\partial T}{\partial x} \right)_{\text{right}} + \left(kA \frac{\partial T}{\partial y} \right)_{\text{bottom}} - \left(kA \frac{\partial T}{\partial y} \right)_{\text{top}} + q_b \Delta V = 0$$

(5.2)

Following the first-order differencing scheme discussed earlier, the temperature gradient at the left and right is discretized as follows:

$$k\Delta y \frac{T_{i-1,j} - T_{i,j}}{\Delta x} - k\Delta y \frac{T_{i,j} - T_{i+1,j}}{\Delta x} + k\Delta x \frac{T_{i,j-1} - T_{i,j}}{\Delta y} - k\Delta x \frac{T_{i,j} - T_{i,j+1}}{\Delta y} + q_b \Delta x \Delta y = 0$$

(5.3)

Simplifying the equation gives

$$kA_x \frac{T_{i-1,j} - 2T_{i,j} + T_{i+1,j}}{\Delta x} + kA_y \frac{T_{i,j-1} - 2T_{i,j} + T_{i,j+1}}{\Delta y} + q_b \Delta x \Delta y = 0 \quad (5.4)$$

where A_x and A_y are the area where heat conduction flux is passing through in x and y direction, respectively. Based on Eq. (5.4), one can notice that governing equation is now discretized into algebraic equation which depends on temperature difference among nodal points (or the temperature stored in the cells of the spreadsheet).

Similar to the one-dimensional case, we could also rearrange the equation in form as following:

$$a_{i,j} T_{i,j} = a_{i+1,j} T_{i+1,j} + a_{i-1,j} T_{i-1,j} + a_{i,j+1} T_{i,j+1} + a_{i,j-1} T_{i,j-1} + S \quad (5.5)$$

where $a_{i,j}$, $a_{i-1,j}$, $a_{i+1,j}$, $a_{i,j+1}$, $a_{i,j-1}$, and S are the coefficients of the equation which could be easily implemented in spreadsheet. Details of the implementation could be demonstrated in the following problem examples.

5.3.2 Two-Dimensional Heat Conduction in a Square Steel Column

We first consider a steady two-dimensional heat conduction through a square steel column. The steel column with a cross-sectional area of 5 cm × 5 cm is used to support the structure a furnace. One side of the steel column is exposed to high temperature of 160°C due to the heat loss from the combustion chamber. The other side is exposed to a constant temperature of 60°C which is adjacent to a recovery heat exchanger (see also in Fig. 5.3). The temperature difference forms a constant temperature gradient cross the horizontal direction. On the other hand, insulations were deployed covering the other two surfaces of the steel column. Heat loss through the two surfaces is negligible. Due to the considerably low-temperature change, thermal properties of the steel column are assumed to be independent to the temperature change. The thermal conductivity and capacity are of 54 W/m K and 0.465 kJ/kg K, respectively. The density of steel column is 7750 kg/m^3.

In the heat transfer viewpoint, due to the adiabatic condition at both top and bottom boundaries. There is no heat conduction in vertical direction. The heat transfer within the steel column can simply be treated as one-dimensional heat transfer. Moreover, as temperature gradient and all thermal properties are constant, temperature distribution within the column exhibits a simple linear relationship. In other words the temperature within the column can be determined by simple interpolation with respected to the temperature gradient as follows:

$$T(x) = 160 - \frac{x}{0.05} \times (160 - 60) \tag{5.6}$$

Obviously, solution of above problem is trivial. Solving the problem with numerical method seems to be over-killed. However, this also set an example for the reader to verify their results.

Back to the numerical method, for the ease of implementation, we should further simplify the governing equation to suit the problem. First of all, a 20 × 20 numerical mesh is adopted to discretize the square column. All the

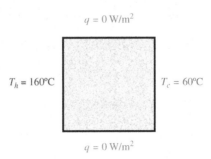

FIGURE 5.3   A schematic of the two-dimensional heat condition in a square steel column.

meshes are therefore in square shape of 2.5 mm × 2.5 mm. In other word, A_x and A_y are identical and Δx is equal to Δy. Moreover, there is no internal heat generation within the column. The governing equation could be simplified as follows:

$$kA\frac{T_{i-1,j} - 2T_{i,j} + T_{i+1,j}}{\Delta x} + kA\frac{T_{i,j-1} - 2T_{i,j} + T_{i,j+1}}{\Delta x} = 0 \qquad (5.7)$$

Dividing the equation by $\frac{kA}{\Delta x}$ and rearrange the equation give

$$T_{i,j} = \frac{T_{i-1,j} + T_{i+1,j} + T_{i,j-1} + T_{i,j+1}}{4} \qquad (5.8)$$

which is simply an arithmetic average of the temperature at the four-neighboring nodal points. One should also notice that this statement is also valid for three-dimensional case. For three-dimensional problem, with square mesh and constant thermal properties, the temperature is the arithmetic average of all 6 four-neighboring nodal temperature. Eq. (5.8) can also be implemented easily in spreadsheet by calculating the average of the four-neighboring cells, as shown in Fig. 5.4. For heat transfer problem with non-uniform meshes, the discretized formulation would be similar to the equation 5.8 while the area and spacing of the mesh are not constant.

In this problem, two types of boundary conditions are adopted in the numerical procedure. Unlike the boundary condition introduced in the

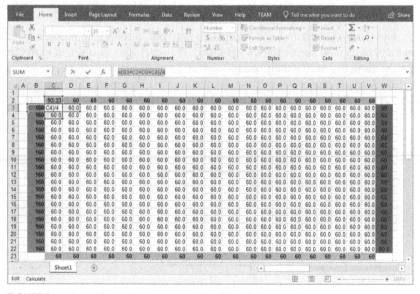

FIGURE 5.4 Spreadsheet showing computation of Eq. (5.8).

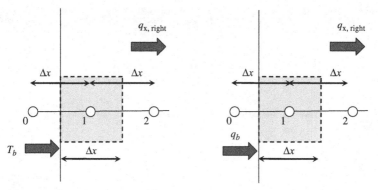

FIGURE 5.5 Mesh arrangement at the constant temperature and heat flux boundary condition with additional imaginary node.

previous chapter, it is more convenient to use additional cells for storing data as the boundary condition.

For the constant temperature, as depicted in Fig. 5.5, temperature is stored in an additional nodal point outside the computational domain (i.e., steel column). One should notice this additional node is imaginary and does not practically exist in the problem. However, since the imaginary node has the same spacing with all internal nodal points, Eq. (5.8) is still valid. Therefore the first nodal point can be determined from the arithmetic average of the four-neighboring nodal points. The only problem left is how to specify a correct temperature in the imaginary node according to the boundary condition.

For the *constant temperature* boundary condition, as discussed in Chapter 4, the heat transfer from the boundary to the first node has only half grid size. In order to maintain the same temperature gradient, temperature at the imaginary node can be determined by extrapolation as follows:

$$kA\frac{T_1 - T_b}{\Delta x/2} = kA\frac{T_1 - T_0}{\Delta x}$$

$$T_0 = T_1 - 2 \times (T_1 - T_b) \tag{5.9}$$

where T_b is the temperature at the boundary (i.e., 160 and 60°C at both sides in this problem).

For the *adiabatic* boundary condition the treatment could even more simple. In physical viewpoint, adiabatic condition has no heat flux passing through the boundary. To set the heat flux as zero, we could simply specify the temperature at the imaginary nodal to be identical to the adjacent node.

$$kA\frac{T_1 - T_0}{\Delta x} = 0$$

$$T_0 = T_1 \tag{5.10}$$

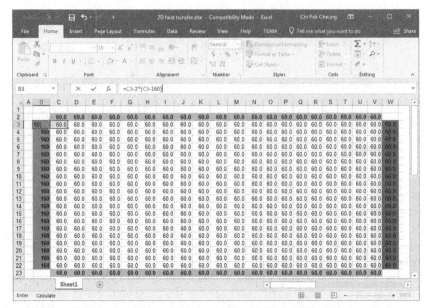

FIGURE 5.6 Constant temperature boundary condition implementation.

Using Eq. (5.10) the boundary condition can be implemented in the additional node. Figs. 5.6 and 5.7 demonstrate how the formulation be implemented as the boundary condition in the spreadsheet.

After implementing the formulation, all the cells are correlated to each other. This also forms a set of simultaneous equations governing all cells with appropriate physical interpretations. The solution can now be obtained by iterative procedures. In most of the spreadsheet, iterative option requires user to enable the iteration functionality. For Microsoft© excel users, the option can be enabled by following the file menu and select option. Under the Formulas tab, user can enable the iterative calculation and specify the maximum iteration number and maximum change as the convergence target as shown in Fig. 5.8.

After enabling the iterative calculation the spreadsheet will carry out the iterative process and obtain a converged solution as shown in Fig. 5.9. Depend on the hardware of the computer, such iterative process normally takes less than a few seconds to complete. Comparing to solving the solution by matrix operation, the iterative method appears to be faster and more convenient. In fact, similar procedure could be also implemented for one-dimensional heat conduction problem. It depends totally on user preference if direct or iterative method is adopted in any given problem.

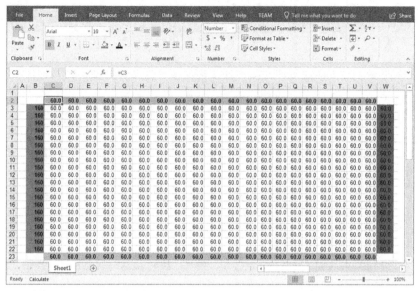

FIGURE 5.7 Adiabatic boundary condition implementation.

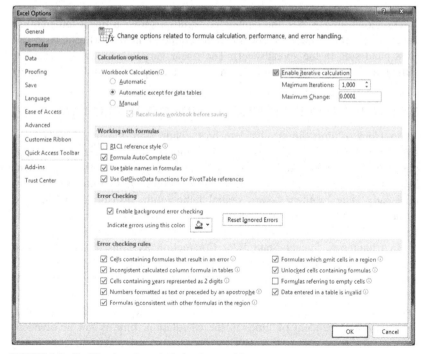

FIGURE 5.8 Enabling iterative calculation in spreadsheet.

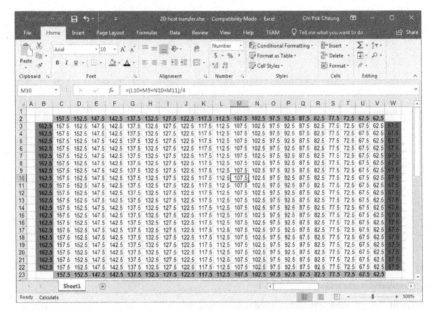

FIGURE 5.9 Converaged solution for the temperature distribution within the square steel column.

As shown in Fig. 5.9, the converged solution shows a linear temperature variation from the high- to the low-temperature boundary (i.e., left to the right hand side in the spreadsheet). The linear variation follows exactly the same temperature gradient with the interpolation expression in Eq. (5.6).

5.4 TIME-DEPENDENT HEAT CONDUCTION PROBLEM

Apart from solving the steady state heat conduction, the transient thermal stress imposes to the system is also a common problem where engineers or designers constantly require a thorough evaluation of the time-dependent heat conduction within the system. The discussion earlier so far focuses only on steady state problem where spatial temperature distribution within the domain is concerned. In this section, we introduce two numerical methods for incorporating time-dependent consideration in the iterative process.

In Chapter 4, Steady State Heat Conduction Systems, finite difference approach is introduced to approximate the spatial temperature gradient and solving the temperature distribution with a set of discrete numerical mesh. The solution obtained represents temperature distribution when the heat conduction process is fully developed and the temperature at any given spatial location is independent of time. In the calculation, time is not involved in the governing equations (as well as the discretized algebraic equations). Nonetheless, for transient heat conduction problem, temperature change with

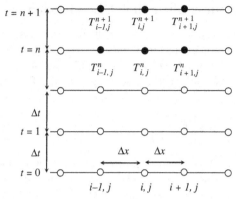

FIGURE 5.10   Notation for the spatial and temporal temperature change in transient problem.

time as well as spatial locations. The discretization procedure should therefore involve temperature derivatives with respect to time and space. In other word the time variation become an additional dimension on top of the spatial discretization as shown in Fig. 5.10. As depicted, temperature distribution is stored in spatial grid which is two-dimensional in our case. To consider time variation, temperature at every nodal point change with each time step (i.e., $t = \Delta t$). At the initial condition (i.e., $t = 0$), temperature at each node equal to the initial temperature. As the heat conduction take place, temperature would rise after a certain time internal or time step. This process will continue with the temperature at each node changing with respect to time until the steady state condition achieved.

For example, instead of steady state heat conduction, the steel column in our previous example has an initial temperature of 60°C. High temperature from the furnace starts heating up the steel column when time equal to zero (i.e., $t = 0$). The additional heat coming through heats up the steel column gradually. After 2 s (i.e., $t = 2$ s), temperature nearby the furnace rise start transferring heat to the other side connected to the heat exchanger. Considering that we measured the temperature at each node for every 2 s time-internal, we would record the temperature development from initial for every 2 s until there is no temperature rise in any nodal point (which is corresponding to the steady state condition).

This example elucidates the discretization nature for transient problem. Instead of solving a continue time-dependent temperature rise, numerical method only calculates the "snapshot" or instantaneous temperature value at a given time for every nodal point. Obviously, adopting a small time-step will reduce the error in general and provide more comprehensive information. Unfortunately, more computational step and resource would require for small time step. Moreover, with time as additional dimension, the notation $T_{i,j}^n$ refers to the temperature at nodal point (i, j) at the time $(t = n \times \Delta t)$.

5.4.1   Implicit and Explicit Time Discretization

Different from the steady state heat conduction, for transient heat conduction, the governing equation should consider the change of energy content at a given finite element. As discussed in Chapter 4, Steady State Heat Conduction Systems, the generic governing equation for transient heat conduction is given as

$$\left[\frac{\partial}{\partial x}\left(k\frac{\partial T}{\partial x}\right) + \frac{\partial}{\partial y}\left(k\frac{\partial T}{\partial y}\right) + \frac{\partial}{\partial z}\left(k\frac{\partial T}{\partial z}\right) + q_b\right] = \frac{\partial(\rho CT)}{\partial t} \qquad (5.11)$$

This generic heat conduction governing equation is applicable for the transient three-dimensional heat conduction problem with internal volumetric heat source. Instead of zero the spatial heat conduction is now equal to the energy change per unit volume of the finite element with respect to time. Using the finite difference method the time-dependent energy change can be approximated as

$$\frac{\partial(\rho CT)}{\partial t} = \rho C\frac{T_{i,j}^{n+1} - T_{i,j}^n}{\Delta t} \qquad (5.12)$$

where $T_{i,j}^n$ and $T_{i,j}^{n+1}$ refer to the temperature at nodal point (i, j) at the times of $t = n \times \Delta t$ and $t = (n + 1) \times \Delta t$, respectively. The physical interpretation of Eq. (5.12) is the temperature change between the time step Δt at the nodal point (i, j) caused by the heat transfer and internal heat source. To complete the discretization, we should now consider what is the dominant factor contributing to the temperature rise during time step.

There are generally two main methods in approaching this matter. One method assumes that the temperature rise is entirely caused by energy input due to the heat transfer and internal heat source in the previous time step $(t = n \times \Delta t)$. In other word, temperature of the nodal point (i, j) at time $t = (n + 1) \times \Delta t$ is determined by its temperature, internal heat generation, and the heat conduction from the neighboring points (i.e., $(i + 1, j)$, $(i - 1, j)$) at previous time step $(t = n \times \Delta t)$, as shown in Fig. 5.11. This is commonly

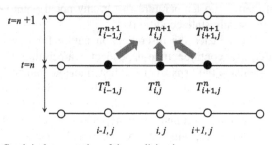

FIGURE 5.11   Graphrical presentation of the explicit scheme.

referred as *explicit method* where the discretization equation for two-dimensional heat conduction problems becomes

$$kA_x \frac{T^n_{i-1,j} - 2T^n_{i,j} + T^n_{i+1,j}}{\Delta x} + kA_y \frac{T^n_{i,j-1} - 2T^n_{i,j} + T^n_{i,j+1}}{\Delta y} + qb^n \Delta V = \rho C \Delta V \frac{T^{n+1}_{i,j} - T^n_{i,j}}{\Delta t}$$

(5.13)

From the formulation, one could notice that the explicit method is a time marching scheme where temperature change in next time step is totally relied on temperature in present. Neighboring nodal points in next time step has no influence at all. The main advantage of such a time marching scheme is the relatively low computational requirement. Based on the given guess or initial condition, temperature at next time step could be determined in a process somewhat analogy to extrapolation from the present values. Nonetheless, It will become more transparent in next section; the explicit method suffers from the limitation of relatively small time step. Nonrealistic large time step could pose stability issue in the time-marching or even causing diverge situation.

The other method adopts an opposite approach by assuming the temperature rise is governed by the temperature distribution at the next time step and temperature value at present time. As shown in Fig. 5.12, temperature at the nodal point (i, j) at next time step depends on the heat transfer from neighboring points (i.e., $(i + 1, j)$, $(i - 1, j)$) at the next time step $t = (n + 1) \times \Delta t$ and the present temperature values. This is commonly referred as *implicit method* where the discretization equation for two-dimensional heat conduction problems becomes

$$kA_x \frac{T^{n+1}_{i-1,j} - 2T^{n+1}_{i,j} + T^{n+1}_{i+1,j}}{\Delta x} + kA_y \frac{T^{n+1}_{i,j-1} - 2T^{n+1}_{i,j} + T^{n+1}_{i,j+1}}{\Delta y}$$
$$+ qb^n \Delta V = \rho C \Delta V \frac{T^{n+1}_{i,j} - T^n_{i,j}}{\Delta t}$$

(5.14)

Based on the formulation, one could notice that majority of temperature variables are the next time step where the exact temperature values are unknown. This is the disadvantage of the implicit method. Unlike explicit

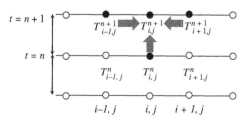

FIGURE 5.12 Graphrical presentation of the implicit scheme.

method where the "future" temperature is extrapolated from the present values, implicit method requires information from the future value which are unknowns in the equation. To obtain these unknown values, it requires to formulate a set of simultaneous equation for all future values at the nodal points. Iterative processes are then adopted to obtain all future value. In other word, implicit method requires iteration process for each time step which obviously is time consuming and computationally heavy and expensive. Nonetheless, as the value is obtained from iteration process at each time step, a larger time step could be used without suffering instability issue as experienced in explicit method.

5.4.2 Stability and Choice of Time Step for Explicit Method

To understand the stability issue, we should firstly understand the formulation in both methods. Considering a transient one-dimensional heat conduction problem (similar to the problem discussed in Chapter 4, Steady State Heat Conduction Systems), the governing equation could then be reduced as follows:

$$\frac{\partial(\rho C T)}{\partial t} = \frac{\partial}{\partial x}\left(k\frac{\partial T}{\partial x}\right) + q_b \tag{5.15}$$

Adopting the finite difference method, the governing equation for any element with a volume of ΔV could be discretized as follows:

$$\rho C \Delta V \frac{T_i^{n+1} - T_i^n}{\Delta t} = kA\frac{T_{i-1} - T_i}{\Delta x} - kA\frac{T_i - T_{i+1}}{\Delta x} + q_b \Delta V \tag{5.16}$$

Notice that the $\Delta V = \Delta x A$, where A is the cross-sectional area of any element. Canceling the area A and rearrange Eq. (5.16) gives

$$\frac{\rho C}{k\Delta t}\Delta x^2(T_i^{n+1} - T_i^n) = (T_{i-1} - 2T_i + T_{i+1}) + q_b\frac{\Delta x^2}{k} \tag{5.17}$$

Introducing a dimensionless number—*mesh Fourier number* as

$$\tau = \frac{k\Delta t}{\rho C \Delta x^2} \tag{5.18}$$

One should notice that Eq. (5.17) is valid for both implicit and explicit method. Depends on the method, present or future temperature values are adopted in the right-hand side. For the *explicit method*, the formulation is expressed as

$$\frac{(T_i^{n+1} - T_i^n)}{\tau} = (T_{i-1}^n - 2T_i^n + T_{i+1}^n) + q_b\frac{\Delta x^2}{k} \tag{5.19}$$

Arranging the equation gives

$$T_i^{n+1} = \tau(T_{i-1}^n + T_{i+1}^n) + (1 - 2\tau)T_i^n + q_b\tau\frac{\Delta x^2}{k} \qquad (5.20)$$

For all internal points, Eq. (5.20) solves the temperature at next time step directly from the present temperature value. This could also consider as an extrapolation of new temperature from the present value. On the other hand, for the *implicit method*, the formulation becomes

$$\frac{(T_i^{n+1} - T_i^n)}{\tau} = (T_{i-1}^{n+1} - 2T_i^{n+1} + T_{i+1}^{n+1}) + q_b\frac{\Delta x^2}{k} \qquad (5.21)$$

Arranging the equation gives

$$(1 + 2\tau)T_i^{n+1} = \tau(T_{i-1}^{n+1} + T_{i+1}^{n+1}) + q_b\tau\frac{\Delta x^2}{k} + T_i^n \qquad (5.22)$$

As discussed, the formulation reveals that implicit method requires iterative process to obtain the new temperature on the right-hand side of the equation. Comparing both methods, we can understand that explicit method is relatively more straight-forward and easier to be implemented in spreadsheet.

Nevertheless, the choice of time step for explicit method must be careful. Otherwise, no realistic result or even divergence situation could occur. As shown in Eq. (5.20), the coefficient for present temperature value (i.e., T_i^n) is $1-2\tau$. From the physical viewpoint the present temperature should have a positive effect on the future temperature. For example, if the heat conduction has resulted a temperature rise in the element, the temperature at next time step is equal to the temperature rise plus previous temperature. Thus the coefficient must be positive to avoid unrealistic calculation. In other word, to maintain stability or realistic explicit calculation, the criteria could be expressed as

$$(1 - 2\tau) \geq 0 \quad \text{or} \quad \tau = \frac{k\Delta t}{\rho C \Delta x^2} \leq \frac{1}{2} \qquad (5.23)$$

With the given thermal property and mesh spacing, one should carefully choose the time step size to avoid instability occur. On the other hand, unrealistic calculation will not occur for the implicit method (see also Eq. (5.22)).

5.4.3   Transient Heat Conduction in a One-Dimensional Steel Bar

To demonstrate the difference between explicit and implicit methods, a one-dimensional transient heat conduction problem is discussed here. In this problem, similar to the two-dimensional problem, we are considering the heat conduction through a steel bar. The steel bar with a 5 cm length is used to support the structure a furnace. One side of the steel bar is exposed to high temperature of 160°C due to the heat loss from the combustion chamber. The other side is exposed to a constant temperature of 60°C which is adjacent to a recovery heat exchanger. All other surfaces of the steel bar are covered by insulation where heat loss is negligible. The thermal conductivity and capacity are of 54 W/m K and 0.465 kJ/kg K, respectively. The density of steel column is 7750 kg/m^3. Initially the steel bar has a uniform temperature of 60°C. We aim to determine the time for the steel bar to achieve steady state temperature distribution using both methods.

Again, although the problem is one-dimensional, readers could notice that the temperature distribution within the steel bar at the steady state condition exhibits as a linear function as follows:

$$T(x) = 160 - \frac{x}{0.05} \times (160 - 60) \tag{5.24}$$

To simplify the calculation, we adopt a total of 10 mesh points to discretize the problem. The mesh spacing equals to 5 cm/10 = 0.5 cm or 0.005 m. For the boundary condition, identical boundary treatment can be adopted directly in this case (see Eq. (5.9)).

With the boundary condition, all internal nodal points can now be evaluated based on the explicit method. Using the *explicit method*, the temperature at the next time step is given by

$$T_i^{n+1} = \tau(T_{i-1}^n + T_{i+1}^n) + (1 - 2\tau)T_i^n \tag{5.25}$$

For the time step, we should first evaluate the maximum allowable time step based on the stability criteria. Using the given thermal properties and the mesh spacing, the stability criteria is evaluated as

$$\Delta t \leq \frac{\rho C \Delta x^2}{2k} = 0.8 \ s \tag{5.26}$$

For simplicity, we could choose the time step as 0.5 s. Afterward, we can now implement Eq. (5.26) for the first node at the first-time step. Details of the implementation in the spreadsheet are shown in Fig. 5.13.

Applying the same formulation for the rest of the mesh nodes, we can obtain the temperature distribution within the steel bar at the first-time step (i.e., $t = 0.5$ s) as shown in Fig. 5.14. One could notice that temperature rise only occurs in the first grid node. This is because only the first grid node is subject to the heat transfer from the boundary condition. Other nodal points have no energy input from neighboring points since all temperatures in previous time step are 60°C.

To obtain the temperature in other time step, one could simply apply formulation to other rows in the spreadsheet. This can be done by simply highlighting the first row of our time step (i.e., $t = 0.5$ s) then drag the AutoFill Handle down to the bottom of the spreadsheet (Fig. 5.15).

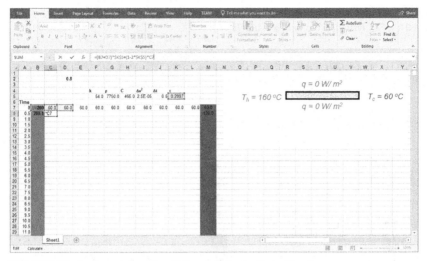

FIGURE 5.13 Implementation of explicit method for the first grid node.

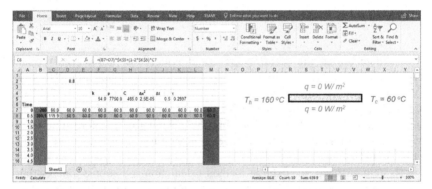

FIGURE 5.14 Applying the explicit method for the whole domain at the first time step.

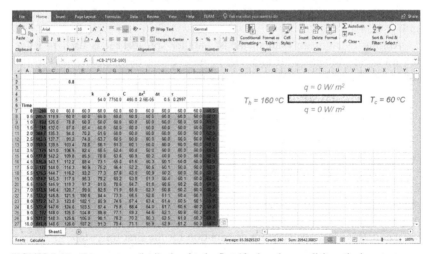

FIGURE 5.15 Temperature distribution for the first 10 s based on explicit method.

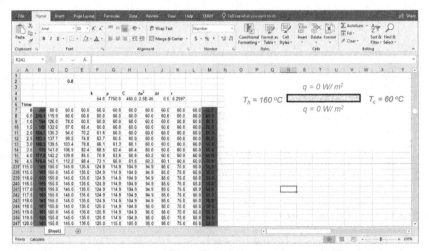

FIGURE 5.16 The steady state temperature distribution predicted by the explicit method.

As depicted in the picture, one could notice that all internal points record a temperature rise during the first 10 s of heat conduction. The rate of such temperature rise depends on the thermal diffusivity of the steel bar. Comparing to our steady state result the temperature distribution is yet to achieve steady state. Further applying the formulation at more time step, we could obtained a steady state temperature distribution at 120 s (Fig. 5.16).

As demonstrated, the explicit method calculation is a simple time marching. With the time step of 0.5 s, it takes 240-time steps to obtain steady state solution. To accelerate the calculation, one could change the time step to be 0.8 s. The same time-distribution would be obtained. However, using a time step higher than 0.8 s (e.g., 1.0 s) would cause a diverged solution as shown in Fig. 5.17.

Aiming to compare both methods, another spreadsheet is used to solve the transient problem with implicit method. Using the *implicit method*, the temperature at the next time step is given by

$$T_i^{n+1} = \frac{\tau}{(1 + 2\tau)}(T_{i-1}^{n+1} + T_{i+1}^{n+1}) + \frac{1}{(1 + 2\tau)}T_i^n \qquad (5.27)$$

For the time step, for the sack of comparison, we choose the time step as 0.5 s (Fig. 5.18).

Similar to the explicit method, we firstly implement the formulation in the first node at the first-time step. As the implicit method requires future values in the calculation, one should be reminded to the enable the iterative process option in the spreadsheet. Applying the formulation to the first row, we could obtain the temperature distribution for the first-time step.

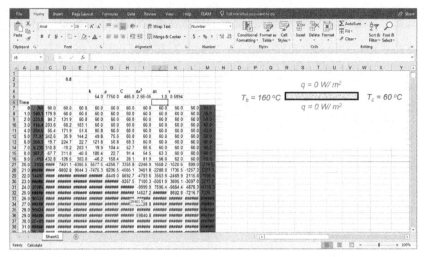

FIGURE 5.17 A divergence problem occurred when 1.0 s time step is employed.

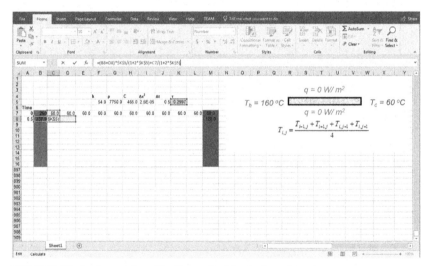

FIGURE 5.18 Implementation of implicit method for the first grid node.

Comparing the result with the explicit method, one can notice that temperature rise occurs not only at the first node but also at three other nodes. The temperature change in other nodes is caused by the heat conduction from the first node which sequentially transfers to other nodes. Such heat conduction effect is neglected in explicit method which imposes some numerical error in each time step (Fig. 5.19).

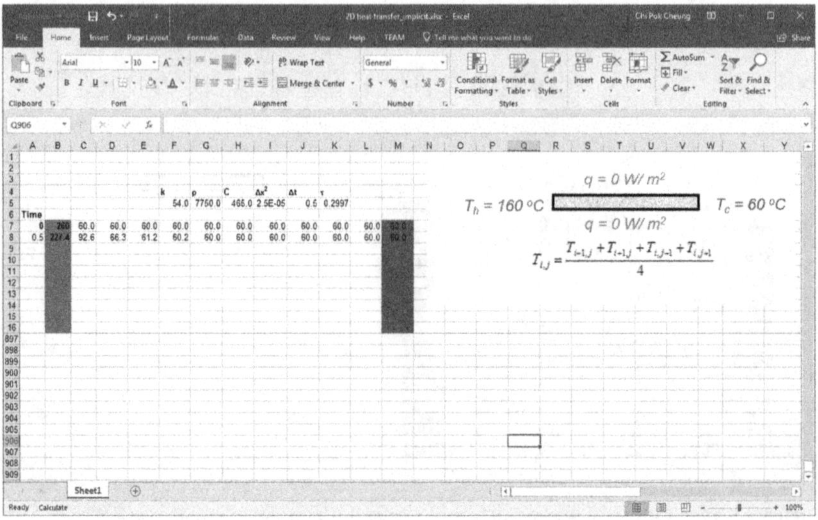

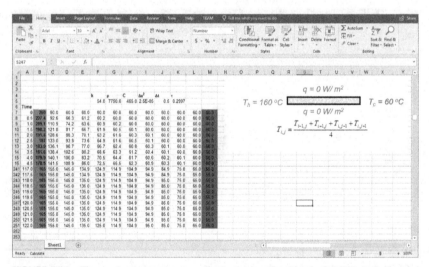

FIGURE 5.19 Applying the implicit method for the whole domain at the first time step.

FIGURE 5.20 The steady state temperature distribution predicted by the implicit method.

As shown in Fig. 5.20, using the implicit method, temperature change at all nodal points achieve the steady state distribution around 122 s which slightly longer than the 120 s predicted by explicit method. In most of the cases, for the same time step, implicit method tends to yield more accurate results compared to the explicit method. This is partly due to the additional

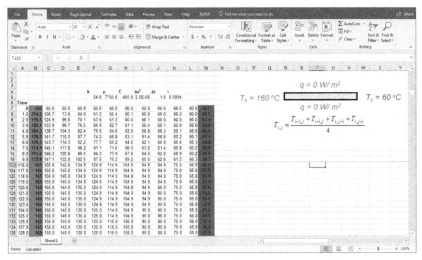

FIGURE 5.21 A divergence problem occurred when 1.0 s time step is employed.

considerations of temperature change in future time-step. Obviously, implicit method requires more computational steps and resources throughout the iterative process.

On the other hand the implicit method is numerically more stable compared to explicit method. The iterative process in the calculation ensures each nodal point is having realistic results. Therefore a larger time step could be employed using implicit method. Fig. 5.21 shows the steady state temperature distribution predicted by the implicit method using a time-step of 1.0 s. There is no divergence problem occurred in this case. The time for achieving steady state is predicted as around 126 s. The different in predictions could be caused by the numerical error in the time discretization. One should realize that we have adopted a first-order differencing scheme for the time discretization for both methods. Thus a larger time step would impose higher discretization and numerical errors in the calculation. Reducing the time step could minimize such error, however, it is at the cost of more computational steps and time.

Chapter 6

Electrical Power Systems

6.1 ELECTRICAL SYSTEMS

An electrical system consists of many different forms of components such as motors, resistors, capacitors, and transistors. These components are designed to be connected in an electrical circuit. The primary objective is to drive the electrical circuit with two electrical characteristics, i.e., voltage and current, which can be measured for understanding and control of the electrical systems, so that desirable system outcomes can be achieved.

Voltage is the difference in electrical potential between two points in space, and in the case of a connected electrical system, two points in the circuit. It is a measure of the amount of energy gained or lost by moving a unit of positive charge from one point to another. Voltage is measured in units of Joules per Coulomb, known as a Volt (V). It is important to remember that voltage is not an absolute quantity. Instead, it is always measured as a value between two points.

On the other hand, electric current is the rate at which electric charge flows through a given area. Current is measured in the unit of Coulombs per second, which is known as an ampere (A).

Three basic electrical component parameters are then derived from these two characteristics. A resistor is an electrical component that restricts the flow of electric current between its end points. Due to this restriction, there is a voltage v between the end points and a measurable current i flowing through the resistor. The resistance R is therefore defined by Eq. (6.1). In practice, the resistance is a characteristic of the materials of the resistor and hence R is a constant.

$$R = \frac{v}{i} \qquad (6.1)$$

A capacitor is an electrical component that stores electric charge. As electric charge flows into the capacitor, it can be measured as a current. As the electric charge builds up in the capacitor, its voltage increases. Hence, the capacitance C of a capacitor is given by Eq. (6.2). The integral term means the total amount of charges built up by current flowing into the capacitor over a defined period of time. The current varies over time but the capacitance is a constant—property of the capacitor material.

Demystifying Numerical Models. DOI: https://doi.org/10.1016/B978-0-08-100975-8.00006-0

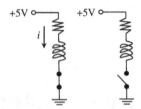

FIGURE 6.1   A motor equivalent circuit model.

$$C = \frac{\int i\, dt}{v} \tag{6.2}$$

An inductor is an electrical component that emits strong electromagnetic force in its surroundings when it is excited by putting a voltage across the inductor. The electromagnetic field induces a back electromagnetic current opposing the excitation voltage. The relationship between the voltage and current for an inductor can be expressed as Eq. (6.3). The parameter L is the inductance of the inductor.

$$v = L\frac{di}{dt} \tag{6.3}$$

These three fundamental electric components can be used to model any electrical systems, for example, a DC motor with a rated voltage of 5 V can be modeled with resistors and transducers connected together as shown in Fig. 6.1. Likewise, motor circuits with soft starters can be modeled with a capacitance in place. Analysis of electrical system is therefore structured around solving responses of the electrical characteristics.

6.2   ANALYSIS OF DC MOTOR CIRCUITS

Motor circuits generally need to be considered together with the mechanical load that the motor is driving. For initial introduction, we can start with analyzing a motor system with no load, i.e., free running shaft. In this case, the motor performance responses are purely dictated by the electrical characteristics of the electrical element.

The direct current (DC) motor circuit in Fig. 6.1 can be described in terms of current i (the dependent variable) and time t (the independent variable). Using Eqs. (6.1) and (6.3), the circuit can be represented by the following differential equation:

$$v = L\frac{di}{dt} + Ri \tag{6.4}$$

If the input voltage v is kept constant, i.e., $v = V$ (where V is a constant), Eq. (6.4) can be re-written in first-order derivative form as shown in Eq. (6.5).

$$\frac{di}{dt} = \frac{V - Ri}{L} \tag{6.5}$$

Using Euler' equation (Eq. (2.21)), and taking an initial current value of i (t_0) at time zero t_0, the current $i(t_1)$ at time t_1 is given by:

$$i(t_1) = i(t_0) + (t - t_0)\frac{di}{dt}\bigg|_{t=t_0} = i(t_0) + (t - t_0)\left(\frac{V - Ri(t_0)}{L}\right) \tag{6.6}$$

At the end of the period, the differential in Eq. (6.5) can then be updated:

$$\frac{di}{dt}\bigg|_{t=t_1} = \frac{V - Ri(t_1)}{L} \tag{6.7}$$

Similarly, the current $i(t_j)$ at any time t_j is given by:

$$i(t_j) = i(t_{j-1}) + (t_j - t_{j-1})\frac{di}{dt}\bigg|_{t=t_{j-1}} \tag{6.8}$$

To illustrate the solution, if the electrical device constants are $L = 10$ H, $V = 5$ V, $R = 3\ \Omega$. The time domain responses of current i can be computed from Eq. (6.8) using numerical integration step $h = 0.5$ s. The circuit is initially not energized, i.e., the switch is off and there is no current flowing. The numerical solution can be presented in Table 6.1 and graphically in Fig. 6.2.

The analytical solution of Eq. (6.5) is:

$$i = \frac{V}{L}\left(1 - e^{-(R/L)t}\right) \tag{6.9}$$

The analytical solution for this problem is also plotted in Fig. 6.2.

Rows 1 and 2 of Table 6.1 are computed as follows. For $t = 0.0$, i.e., row 1, the initial current and its derivatives are zero. For $t = 0.5$, i.e., row 2, substitute t_0 values into Eq. (6.6),

$$i(0.5) = i(t_0) + (t - t_0)\frac{di}{dt}\bigg|_{t=t_0} = i(0.0) + 0.5\left(\frac{5 - 3 \times 0}{10}\right) = 0.25 \tag{6.10}$$

Similarly, for $t = 1.0$,

$$i(1.0) = i(t_0) + (t - t_0)\frac{di}{dt}\bigg|_{t=t_0} = i(0.5) + 0.5\left(\frac{5 - 3 \times 0.5}{10}\right) = 0.4625 \tag{6.11}$$

Other rows are calculated similarly.

TABLE 6.1 Numerical Solution of DC Motor Only Circuit Responses

t	$\frac{di}{dt}$	i(t)	i(t) Analytical	t	$\frac{di}{dt}$	i(t)	i(t) Analytical
0.0	0.0	0.0	0.0000	8.0	0.0437	1.5429	1.5155
0.5	0.5	0.25	0.2322	8.5	0.0371	1.5615	1.5365
1.0	0.425	0.4625	0.4320	9.0	0.0316	1.5773	1.5547
1.5	0.3613	0.6431	0.6040	9.5	0.0268	1.5907	1.5703
2.0	0.3071	0.7967	0.7520	10.0	0.0228	1.6021	1.5837
2.5	0.2610	0.9272	0.8794	10.5	0.0194	1.6118	1.5952
3.0	0.2219	1.0381	0.9891	11.0	0.0165	1.6200	1.6052
3.5	0.1886	1.1324	1.0834	11.5	0.0140	1.6270	1.6138
4.0	0.1603	1.2125	1.1647	12.0	0.0119	1.6329	1.6211
4.5	0.1362	1.2806	1.2346	12.5	0.0101	1.6380	1.6275
5.0	0.1158	1.3385	1.2948	13.0	0.0086	1.6423	1.6329
5.5	0.0984	1.3878	1.3466	13.5	0.0073	1.6460	1.6376
6.0	0.0837	1.4296	1.3912	14.0	0.0062	1.6491	1.6417
6.5	0.0711	1.4652	1.4295	14.5	0.0053	1.6517	1.6452
7.0	0.0605	1.4954	1.4626	15.0	0.0045	1.6539	1.6482
7.5	0.0514	1.5211	1.4910				

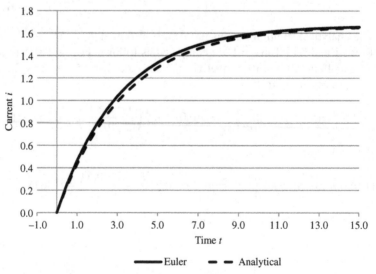

FIGURE 6.2 Graphical responses—both Euler and analytical solutions.

6.3   ANALYSIS OF RLC CIRCUIT

RLC circuits are electrical systems consisting of resistors (R), inductors (L), and capacitors (C). They have both differentiating and integrating terms. Since numerical integration methods do not have a mix of differential and integral terms, formulation of the numerical integration table is done with higher levels derivatives. The Runge–Kutta method is commonly used to find an accurate numerical solution. The process of applying Runge–Kutta method is illustrated with examples.

Using the fundamental electrical system principles, any electrical system can be represented by a network of basic electrical elements as shown in Fig. 6.3. The values are $R = 60\ \Omega$, $L = 200$ mH, $C = 100\ \mu$F.

The voltage $v_b(t)$ can be represented by the voltage across any element:

$$v_b = R i_R = L \frac{d i_L}{dt} = \frac{1}{C} \int i_C\ dt \qquad (6.12)$$

The total current is driven by the constant current source:

$$i = 5\ \sin(\omega t) \qquad (6.13)$$

where $\omega = 1$. The initial condition is $v_b(0) = 0$, then

$$i = i_R + i_L + i_C \qquad (6.14)$$

Combining Eqs. (6.12) and (6.13),

$$i = \frac{v_b}{R} + \int \frac{v_b}{L} dt + C \frac{d v_b}{dt} \qquad (6.15)$$

Differentiate Eq. (6.14) to eliminate the integration term and combine with Eq. (6.13),

$$\frac{1}{R} \frac{d v_b}{dt} + \frac{v_b}{L} + C \frac{d^2 v_b}{dt^2} - 5\omega\ \cos \omega t = 0 \qquad (6.16)$$

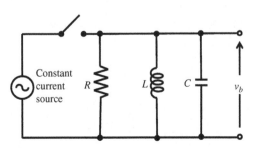

FIGURE 6.3   Electrical network with all three basic electrical elements.

Let $v_1 = \dfrac{dv_b}{dt}$ and noting $v_1(0) = 0$.

Eq. (6.16) can be written as two simultaneous differential equations:

$$\frac{dv_1}{dt} = \frac{5\omega \cos \omega t}{C} - \frac{v_b}{CL} - \frac{v_1}{CR}$$

$$\frac{dv_b}{dt} = v_1 \tag{6.17}$$

This example is analyzed by Runge−Kutta method fourth order. At $t = 0$,

$$\frac{dv_1}{dt}(0) = \frac{5\omega \cos \omega \times 0}{C} - \frac{v_b(0)}{CL} - \frac{v_1(0)}{CR} = 0.05$$

$$\frac{dv_b}{dt}(0) = v_1(0) = 0.0 \tag{6.18}$$

With an integration step of $h = 0.1$, the value of v_b and v_1 at $t = 0 + h/2$ are:

$$v_1\left(\frac{h}{2}\right) = v_1(0) + \frac{h}{2}\frac{dv_1}{dt}(0) = 0.0025$$

$$v_b\left(\frac{h}{2}\right) = v_b(0) + \frac{h}{2}\frac{dv_b}{dt}(0) = 0.0 \tag{6.19}$$

The slope at the mid-interval is then evaluated from the new v values:

$$\left.\frac{dv_1}{dt}\right|_{0+\frac{h}{2},\#2} = \frac{5\omega \cos \omega \times \left(\frac{h}{2}\right)}{C} - \frac{v_b\left(\frac{h}{2}\right)}{CL} - \frac{v_1\left(\frac{h}{2}\right)}{CR} = 0.049937$$

$$\left.\frac{dv_b}{dt}\right|_{0+\frac{h}{2},\#2} = v_1\left(\frac{h}{2}\right) = 0.0025 \tag{6.20}$$

The estimate for the v values are then revised at the mid interval, $t = 0 + h/2$:

$$v_1\left(\frac{h}{2}\right) = v_1(0) + \left.\frac{h}{2}\frac{dv_1}{dt}\right|_{0+\frac{h}{2},\#2} = 0.002497$$

$$v_b\left(\frac{h}{2}\right) = v_b(0) + \left.\frac{h}{2}\frac{dv_b}{dt}\right|_{0+\frac{h}{2},\#2} = 0.000125 \tag{6.21}$$

With the new v values, the revised slope at the mid interval, $t = 0 + h/2$:

$$\left.\frac{dv_1}{dt}\right|_{0+\frac{h}{2},\#3} = \frac{5\omega\cos\omega \times \left(\frac{h}{2}\right)}{C} - \frac{0.00025}{CL} - \frac{0.005}{CR} = 0.049937$$

(6.22)

$$\left.\frac{dv_b}{dt}\right|_{0+\frac{h}{2},\#3} = v_1\left(\frac{h}{2}\right) = 0.002497$$

The estimate at the end of interval, $t = 0 + h$:

$$v_1(h) = v_1(0) + h\left.\frac{dv_1}{dt}\right|_{0+\frac{h}{2},\#3} = 0.004994$$

(6.23)

$$v_b(h) = v_b(0) + h\left.\frac{dv_b}{dt}\right|_{0+\frac{h}{2},\#3} = 0.00025$$

Using the end of interval v values, estimate the slope at end of interval:

$$\left.\frac{dv_1}{dt}\right|_{0+h} = \frac{5\omega\cos\omega \times (h)}{C} - \frac{0.000497}{CL} - \frac{0.00995}{CR} = 0.049749$$

(6.24)

$$\left.\frac{dv_b}{dt}\right|_{0+h} = v_1(h) = 0.004994$$

Finally for this integration step, the v values estimated at the end of interval is given by:

$$v_1(h) = v_1(0) + \frac{h}{6}\left(\left.\frac{dv_1}{dx}\right|_0 + 2\left.\frac{dv_1}{dx}\right|_{0+\frac{h}{2},\#2} + 2\left.\frac{dv_1}{dx}\right|_{0+\frac{h}{2},\#3} + \left.\frac{dv_1}{dx}\right|_{0+h}\right) = 0.004992$$

$$v_b(h) = v_b(0) + \frac{h}{6}\left(\left.\frac{dv_b}{dx}\right|_0 + 2\left.\frac{dv_b}{dx}\right|_{0+\frac{h}{2},\#2} + 2\left.\frac{dv_b}{dx}\right|_{0+\frac{h}{2},\#3} + \left.\frac{dv_b}{dx}\right|_{0+h}\right) = 0.000250$$

(6.25)

The computed values in Eqs. (6.18) to (6.25) are entered to row $t = 0.0$ in Table 6.2. Note that numbers are truncated to four decimal places in Table 6.2.

TABLE 6.2 Numerical Solution of a General DC Circuit (Rows Shown to $t_0 = 2.0$ Only. The Higher t_0 Rows Are Computed Similarly)

Eqs.		(6.13)		(6.14)		(6.15)		(6.16)		(6.17)		(6.18)		(6.19)		(6.20)		(6.21)	
t_0	$v_1(t_0)$	$v_b(t_0)$	$\frac{dv_1}{dt}(0)$	$\frac{dv_b}{dt}(0)$	$v_1(t_0+\frac{h}{2})$	$v_b(t_0+\frac{h}{2})$	$\frac{dv_1}{dt}\big|_{0+\frac{h}{2},g2}$	$\frac{dv_b}{dt}\big|_{0+\frac{h}{2},g2}$	$v_1(t_0+\frac{h}{2})$	$..$	$\frac{dv_1}{dt}\big|_{0+\frac{h}{2},g3}$	$\frac{dv_b}{dt}\big|_{0+\frac{h}{2},g3}$	$v_1(t_0+h)$	$v_b(t_0+h)$	$\frac{dv_1}{dt}\big|_{0+h}$	$\frac{dv_b}{dt}\big|_{0+h}$	$v_1(t_0+h)$	$v_b(t_0+h)$
0.0	0.0000	0.0000	0.0500	0.0000	0.0025	0.0000	0.0499	0.0025	0.0025	0.0001	0.0499	0.0025	0.0050	0.0002	0.0497	0.0050	0.0050	0.0002
0.1	0.0050	0.0002	0.0497	0.0050	0.0075	0.0005	0.0494	0.0075	0.0075	0.0006	0.0494	0.0075	0.0099	0.0010	0.0490	0.0099	0.0099	0.0010
0.2	0.0099	0.0010	0.0490	0.0099	0.0124	0.0015	0.0484	0.0124	0.0124	0.0016	0.0484	0.0124	0.0148	0.0022	0.0478	0.0148	0.0148	0.0022
0.3	0.0148	0.0022	0.0478	0.0148	0.0172	0.0030	0.0470	0.0172	0.0171	0.0031	0.0470	0.0171	0.0195	0.0039	0.0460	0.0195	0.0195	0.0039
0.4	0.0195	0.0039	0.0460	0.0195	0.0218	0.0049	0.0450	0.0218	0.0217	0.0050	0.0450	0.0217	0.0240	0.0061	0.0439	0.0240	0.0240	0.0061
0.5	0.0240	0.0061	0.0439	0.0240	0.0262	0.0073	0.0426	0.0262	0.0261	0.0074	0.0426	0.0261	0.0282	0.0087	0.0413	0.0282	0.0282	0.0087
0.6	0.0282	0.0087	0.0413	0.0282	0.0303	0.0101	0.0398	0.0303	0.0302	0.0102	0.0398	0.0302	0.0322	0.0118	0.0382	0.0322	0.0322	0.0118
0.7	0.0322	0.0118	0.0382	0.0322	0.0341	0.0134	0.0366	0.0341	0.0340	0.0135	0.0366	0.0340	0.0359	0.0152	0.0348	0.0359	0.0359	0.0152
0.8	0.0359	0.0152	0.0348	0.0359	0.0376	0.0170	0.0330	0.0376	0.0375	0.0170	0.0330	0.0375	0.0392	0.0189	0.0311	0.0392	0.0392	0.0189
0.9	0.0392	0.0189	0.0311	0.0392	0.0407	0.0209	0.0291	0.0407	0.0406	0.0210	0.0291	0.0406	0.0421	0.0230	0.0270	0.0421	0.0421	0.0230
1.0	0.0421	0.0230	0.0270	0.0421	0.0434	0.0251	0.0249	0.0434	0.0433	0.0252	0.0249	0.0433	0.0446	0.0273	0.0227	0.0446	0.0446	0.0273
1.1	0.0446	0.0273	0.0227	0.0446	0.0457	0.0295	0.0204	0.0457	0.0456	0.0296	0.0204	0.0456	0.0466	0.0319	0.0181	0.0466	0.0466	0.0319
1.2	0.0466	0.0319	0.0181	0.0466	0.0475	0.0342	0.0158	0.0475	0.0474	0.0343	0.0158	0.0474	0.0482	0.0366	0.0134	0.0482	0.0482	0.0366
1.3	0.0482	0.0366	0.0134	0.0482	0.0488	0.0390	0.0109	0.0488	0.0487	0.0391	0.0109	0.0487	0.0493	0.0415	0.0085	0.0493	0.0493	0.0415
1.4	0.0493	0.0415	0.0085	0.0493	0.0497	0.0440	0.0060	0.0497	0.0496	0.0440	0.0060	0.0496	0.0499	0.0465	0.0035	0.0499	0.0499	0.0465
1.5	0.0499	0.0465	0.0035	0.0499	0.0500	0.0490	0.0010	0.0500	0.0499	0.0490	0.0010	0.0499	0.0500	0.0515	-0.0015	0.0500	0.0500	0.0515
1.6	0.0500	0.0515	-0.0015	0.0500	0.0499	0.0540	-0.0040	0.0499	0.0498	0.0539	-0.0040	0.0498	0.0496	0.0564	-0.0065	0.0496	0.0496	0.0564
1.7	0.0496	0.0564	-0.0065	0.0496	0.0492	0.0589	-0.0089	0.0492	0.0491	0.0589	-0.0089	0.0491	0.0487	0.0613	-0.0114	0.0487	0.0487	0.0614
1.8	0.0487	0.0614	-0.0114	0.0487	0.0481	0.0638	-0.0138	0.0481	0.0480	0.0638	-0.0138	0.0480	0.0473	0.0662	-0.0162	0.0473	0.0473	0.0662
1.9	0.0473	0.0662	-0.0162	0.0473	0.0465	0.0685	-0.0185	0.0465	0.0464	0.0685	-0.0185	0.0464	0.0454	0.0708	-0.0208	0.0454	0.0455	0.0708
2.0	0.0455	0.0708	-0.0208	0.0455	0.0444	0.0731	-0.0231	0.0444	0.0443	0.0730	-0.0231	0.0443	0.0431	0.0752	-0.0253	0.0431	0.0431	0.0752

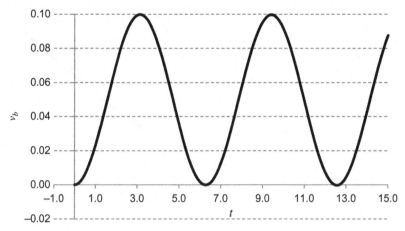

FIGURE 6.4 Plotted data from Table 6.2 (extended to $t = 15.0$).

The next integration interval $t \in [h, 2h)$ is then computed similarly by replacing $t_0 = h$ and the v values in Eq. (6.25). The computation table is presented in Table 6.2.

The data are plotted in Fig. 6.4 for an extended period to $t = 15.0$.

For this example, an analytical solution for v_b exists by solving with Laplace transform:

$$
\begin{aligned}
v_b(t) = {} & 5.0004167 \times 10^{-2} \times e^{-8.333 \times 10^{-5}t}\cos 0.0091283t \\
& + \frac{-4.16754 \times 10^{-6}}{0.0091283} \times e^{-8.333 \times 10^{-5}t}\sin 0.0091283t \qquad (6.26) \\
& - 0.050004167\cos t + 8.33472 \times 10^{-6}\sin t
\end{aligned}
$$

Fig. 6.5 shows the error (magnified 1000 times). It can be seen that although the errors are small, they do accumulate over a period of time. This is inevitable and hence periodic correction by actual measured data is necessary.

6.4 MOTOR DRIVEN POSITION CONTROL SYSTEM

A motor-driven position system consists of three main components: motor, mechanical gearbox and load, position feedback. The schematic diagram of the system is shown in Fig. 6.6.

The electrical motor is represented by an LR circuit using Eq. (6.4). The torque produced by the motor is proportional to the current, where K_m is the motor constant.

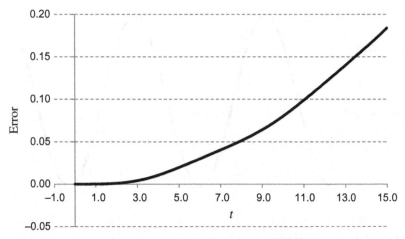

FIGURE 6.5   Comparison of numerical and analytical solutions (error magnified 1000 times).

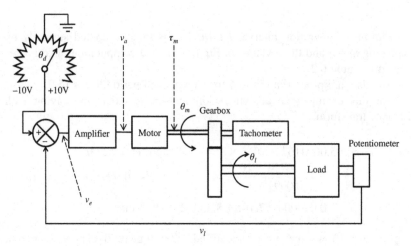

FIGURE 6.6   Schematic diagram of motor-driven position control system.

$$\tau_m = K_m i \tag{6.27}$$

Hence,

$$v_a = L\frac{di}{dt} + Ri \tag{6.28}$$

$$\frac{di}{dt} = \frac{v_a - Ri}{L} \tag{6.29}$$

The mechanical gearbox and load can be represented diagrammatically in Fig. 6.7 incorporating the rotor of the motor, the gears, and the load.

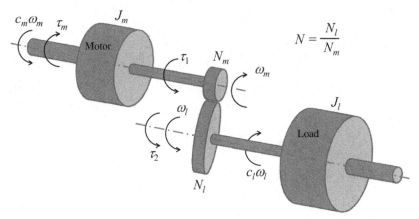

FIGURE 6.7 Diagrammatic representation of the mechanical parts of the system.

Applying free body diagram to the load in Fig. 6.7, the motor shaft motion is:

$$J_m \frac{d\omega_m}{dt} = \tau_m - \tau_1 - c_m \omega_m \qquad (6.30)$$

The load shaft motion is:

$$J_l \frac{d\omega_l}{dt} = \tau_2 - c_l \omega_l \qquad (6.31)$$

where c_m and c_l are the viscous friction coefficients due to bearing motions. Re-arranging terms:

$$\tau_2 = J_l \frac{d\omega_l}{dt} + c_l \omega_l \qquad (6.32)$$

The viscous frictional torque is proportional to the speed of rotation of the shaft (motor or load). At the gearbox,

$$\frac{\omega_m}{\omega_l} = N \qquad (6.33)$$

Similarly,

$$\frac{\tau_2}{\tau_1} = N \qquad (6.34)$$

Combining Eqs. (6.30) to (6.34),

$$NJ_m \frac{d\omega_l}{dt} = \tau_m - \frac{\tau_2}{N} - Nc_m \omega_l \qquad (6.35)$$

$$N^2 J_m \frac{d\omega_l}{dt} = N\tau_m - J_l \frac{d\omega_l}{dt} - c_l \omega_l - N^2 c_m \omega_l \qquad (6.36)$$

Re-arranging,

$$\frac{d\omega_l}{dt}\left(NJ_m + \frac{J_1}{N}\right) + \omega_l\left(Nc_m + \frac{c_l}{N}\right) = \tau_m \tag{6.37}$$

$$J\frac{d\omega_l}{dt} + c\omega_l = \tau_m \tag{6.38}$$

where

$$J = NJ_m + \frac{J_1}{N} \quad \text{and} \quad c = Nc_m + \frac{c_l}{N}$$

Combining with Eq. (6.27),

$$J\frac{d\omega_l}{dt} + c\omega_l = K_m i \tag{6.39}$$

$$\frac{d\omega_l}{dt} = \frac{K_m i - c\omega_l}{J} \tag{6.40}$$

The angular velocity of the load is given by:

$$\frac{d\theta_l}{dt} = \omega_l \tag{6.41}$$

The potentiometer converts angle to voltage:

$$v_d = K_p\theta_d \tag{6.42}$$

$$v_l = K_p\theta_l \tag{6.43}$$

Hence,

$$v_e = K_p(\theta_d - \theta_l) \tag{6.44}$$

The voltage amplifier:

$$v_a = K_a v_e = K_a K_p(\theta_d - \theta_l) \tag{6.45}$$

Now, lets consider an example with constants:

$c = 0.05$ N-m s/rad
$J = 0.0125$ N-m s^2/rad
$K_m = 13.5$ N-m/V
$K_p = 0.02$ V/rad
$K_a = 6$ V/V
$N = 30$
$R = 30$ Ω
$L = 50$ mH

The initial values are:

$$\begin{aligned} \theta_l(0) &= 0 \\ \omega_l(0) &= 0 \\ i(0) &= 0 \end{aligned} \tag{6.46}$$

$\theta_d = \Delta = 1$, i.e., a step input

Combining Eq. (6.29) with Eq. (6.45), and together with Eqs. (6.40) and (6.41), we have three simultaneous differential equations:

$$\frac{di}{dt} = \frac{K_a K_p (\theta_d - \theta_l) - Ri}{L}$$

$$\frac{d\omega_l}{dt} = \frac{K_m i - c\omega_l}{J} \tag{6.47}$$

$$\frac{d\theta_l}{dt} = \omega_l$$

This example is analyzed by Runge–Kutta method fourth order. At $t = 0$,

$$\frac{di}{dt}(0) = \frac{K_a K_p [\Delta - \theta_l(0)] - Ri(0)}{L} = \frac{K_a K_p \Delta}{L} = 0.0024$$

$$\frac{d\omega_l}{dt}(0) = \frac{K_m i(0) - c\omega_l(0)}{J} = 0.0 \tag{6.48}$$

$$\frac{d\theta_l}{dt}(0) = \omega_l(0) = 0.0$$

With an integration step of $h = 0.1$, the value of i, ω_l and θ_l at $t = 0 + h/2$ are:

$$i\left(\frac{h}{2}\right) = i(0) + \frac{h}{2}\frac{di}{dt}(0) = 0.00012$$

$$\omega_l\left(\frac{h}{2}\right) = \omega_l(0) + \frac{h}{2}\frac{d\omega_l}{dt}(0) = 0.0 \tag{6.49}$$

$$\theta_l\left(\frac{h}{2}\right) = \theta_l(0) + \frac{h}{2}\frac{d\theta_l}{dt}(0) = 0.0$$

The slope at the mid-interval is then evaluated from the new i, ω_l, and θ_l values:

$$\left.\frac{di}{dt}\right|_{0+\frac{h}{2},\#2} = \frac{K_a K_p \left[\Delta - \theta_l\left(\frac{h}{2}\right)\right] - Ri\left(\frac{h}{2}\right)}{L} = 0.002328$$

$$\left.\frac{d\omega_l}{dt}\right|_{0+\frac{h}{2},\#2} = \frac{K_m i\left(\frac{h}{2}\right) - c\omega_l\left(\frac{h}{2}\right)}{J} = 0.1296 \tag{6.50}$$

$$\left.\frac{d\theta_l}{dt}\right|_{0+\frac{h}{2},\#2} = \omega_l\left(\frac{h}{2}\right) = 0.0$$

The estimate for the i, ω_l, and θ_l values are then revised at the mid interval, $t = 0 + h/2$:

$$i\left(\frac{h}{2}\right) = i(0) + \frac{h}{2}\frac{di}{dt}\bigg|_{0+\frac{h}{2},\#2} = 0.000116$$

$$\omega_l\left(\frac{h}{2}\right) = \omega_l(0) + \frac{h}{2}\frac{d\omega_l}{dt}\bigg|_{0+\frac{h}{2},\#2} = 0.006480 \qquad (6.51)$$

$$\theta_l\left(\frac{h}{2}\right) = \theta_l(0) + \frac{h}{2}\frac{d\theta_l}{dt}\bigg|_{0+\frac{h}{2},\#2} = 0.0$$

With the new i, ω_l, and θ_l values, the revised slope at the mid interval, $t = 0 + h/2$:

$$\frac{di}{dt}\bigg|_{0+\frac{h}{2},\#3} = \frac{K_aK_p\left[\Delta - \theta_l\left(\frac{h}{2}\right)\right] - Ri\left(\frac{h}{2}\right)}{L} = 0.002330$$

$$\frac{d\omega_l}{dt}\bigg|_{0+\frac{h}{2},\#3} = \frac{K_mi\left(\frac{h}{2}\right) - c\omega_l\left(\frac{h}{2}\right)}{J} = 0.099792 \qquad (6.52)$$

$$\frac{d\theta_l}{dt}\bigg|_{0+\frac{h}{2},\#3} = \omega_l\left(\frac{h}{2}\right) = 0.006480$$

The estimate at the end of interval, $t = 0 + h$:

$$i(h) = i(0) + h\frac{di}{dt}\bigg|_{0+\frac{h}{2},\#3} = 0.000233$$

$$\omega_l(h) = \omega_l(0) + h\frac{d\omega_l}{dt}\bigg|_{0+\frac{h}{2},\#3} = 0.009979 \qquad (6.53)$$

$$\theta_l(h) = \theta_l(0) + h\frac{d\theta_l}{dt}\bigg|_{0+\frac{h}{2},\#3} = 0.000648$$

Using the end of interval i, ω_l, and θ_l values, estimate the slope at end of interval:

$$\frac{di}{dt}\bigg|_{0+h} = \frac{K_aK_p[\Delta - \theta_l(h)] - Ri(h)}{L} = 0.002259$$

$$\frac{d\omega_l}{dt}\bigg|_{0+h} = \frac{K_m i(h) - c\omega_l(h)}{J} = 0.211740 \qquad (6.54)$$

$$\frac{d\theta_l}{dt}\bigg|_{0+h} = \omega_l(h) = 0.009979$$

Finally for this integration step, the i, ω_l, and θ_l values estimated at the end of interval is given by:

$$i(h) = i(0) + \frac{h}{6}\left(\frac{di}{dx}\bigg|_0 + 2\frac{di}{dx}\bigg|_{0+\frac{h}{2},\#2} + 2\frac{di}{dx}\bigg|_{0+\frac{h}{2},\#3} + \frac{di}{dx}\bigg|_{0+h}\right) = 0.000233$$

$$\omega_l(h) = \omega_l(0) + \frac{h}{6}\left(\frac{d\omega_l}{dx}\bigg|_0 + 2\frac{d\omega_l}{dx}\bigg|_{0+\frac{h}{2},\#2} + 2\frac{d\omega_l}{dx}\bigg|_{0+\frac{h}{2},\#3} + \frac{d\omega_l}{dx}\bigg|_{0+h}\right) = 0.011175$$

$$\theta_l(h) = \theta_l(0) + \frac{h}{6}\left(\frac{d\theta_l}{dx}\bigg|_0 + 2\frac{d\theta_l}{dx}\bigg|_{0+\frac{h}{2},\#2} + 2\frac{d\theta_l}{dx}\bigg|_{0+\frac{h}{2},\#3} + \frac{d\theta_l}{dx}\bigg|_{0+h}\right) = 0.000382$$

$$(6.55)$$

The computed values from Eqs. (6.48) to (6.55) are entered to the row $t = 0.0$ in Table 6.3.

The next integration interval $t \in [h, 2\,h)$ is then computed similarly by replacing $t_0 = h$ and the i, ω_l, and θ_l values in Eq. (6.54). Note that Table 6.3 (a) and (b) should be read as the left and right parts of one table.

The data are plotted in Fig. 6.8 for an extended period to $t = 40.0$.

For this example, an analytical solution for v_b exists by solving with Laplace transform:

$$\theta_l = 2.592 \times \begin{bmatrix} 0.387536 - 0.01493 \times e^{-4.15515t} - 0.3726 \times e^{-0.21312t}\cos0.758679t \\ -0.186458 \times e^{-0.21312t}\sin0.758679t \end{bmatrix}$$

$$(6.56)$$

Fig. 6.9 shows the error (magnified 1000 times). It can be seen that due to the oscillatory response, the errors are also oscillatory around the 0.0001 level.

TABLE 6.3 Numerical Solution of a Motor Driven Positional Control System (Rows Shown to $t_0 = 2.0$ Only. The Higher t_0 Rows Are Computed Similarly)

(a) Left-Hand Side Table

Eqs.		(6.43)				(6.44)		(6.45)				(6.46)			(6.47)		
t	$i(t_0)$	$\omega_i(t_0)$	$\theta_i(t_0)$	$\frac{di}{dt}(t_0)$	$\frac{d\omega_i}{dt}(t_0)$	$\frac{d\theta_i}{dt}(t_0)$	$i(t_0+h/2)$	$\omega_i(t_0+h/2)$	$\theta_i(t_0+h/2)$	$\frac{di}{dt}\big	_{0+\frac{1}{2}g}$	$\frac{d\omega_i}{dt}\big	_{0+\frac{1}{2}g}$	$\frac{d\theta_i}{dt}\big	_{0+\frac{1}{2}g}$	$i(t_0+h/2)$	$\omega_i(t_0+h/2)$	$\theta_i(t_0+h/2)$
0.0	0.0000	0.0000	0.0000	0.0024	0.0000	0.0000	0.0001	0.0000	0.0000	0.0023	0.1296	0.0000	0.0001	0.0065	0.0000
0.1	0.0002	0.0112	0.0004	0.0023	0.2068	0.0112	0.0003	0.0215	0.0009	0.0022	0.2875	0.0215	0.0003	0.0255	0.0015
0.2	0.0005	0.0387	0.0028	0.0021	0.3332	0.0387	0.0006	0.0554	0.0047	0.0021	0.3812	0.0554	0.0006	0.0578	0.0055
0.3	0.0007	0.0761	0.0084	0.0020	0.4057	0.0761	0.0008	0.0963	0.0122	0.0019	0.4318	0.0963	0.0008	0.0976	0.0133
0.4	0.0008	0.1187	0.0182	0.0018	0.4421	0.1187	0.0009	0.1408	0.0241	0.0018	0.4534	0.1408	0.0009	0.1414	0.0252
0.5	0.0010	0.1637	0.0323	0.0017	0.4541	0.1637	0.0011	0.1864	0.0404	0.0016	0.4554	0.1864	0.0011	0.1864	0.0416
0.6	0.0012	0.2090	0.0509	0.0016	0.4495	0.2090	0.0013	0.2315	0.0613	0.0015	0.4441	0.2315	0.0013	0.2312	0.0625
0.7	0.0013	0.2532	0.0740	0.0014	0.4337	0.2532	0.0014	0.2749	0.0867	0.0013	0.4235	0.2749	0.0014	0.2744	0.0878
0.8	0.0015	0.2955	0.1015	0.0013	0.4100	0.2955	0.0015	0.3160	0.1162	0.0012	0.3967	0.3160	0.0015	0.3153	0.1173
0.9	0.0016	0.3350	0.1330	0.0011	0.3811	0.3350	0.0016	0.3541	0.1498	0.0011	0.3656	0.3541	0.0016	0.3533	0.1507
1.0	0.0017	0.3715	0.1684	0.0010	0.3485	0.3715	0.0017	0.3890	0.1870	0.0009	0.3316	0.3890	0.0017	0.3881	0.1878
1.1	0.0018	0.4047	0.2072	0.0008	0.3135	0.4047	0.0018	0.4203	0.2275	0.0008	0.2956	0.4203	0.0018	0.4194	0.2282
1.2	0.0019	0.4342	0.2492	0.0007	0.2770	0.4342	0.0019	0.4481	0.2709	0.0006	0.2585	0.4481	0.0019	0.4471	0.2716
1.3	0.0019	0.4600	0.2939	0.0005	0.2397	0.4600	0.0020	0.4720	0.3169	0.0005	0.2208	0.4720	0.0019	0.4711	0.3175
1.4	0.0020	0.4821	0.3411	0.0004	0.2019	0.4821	0.0020	0.4922	0.3652	0.0003	0.1830	0.4922	0.0020	0.4913	0.3657
1.5	0.0020	0.5004	0.3902	0.0003	0.1642	0.5004	0.0020	0.5086	0.4153	0.0002	0.1454	0.5086	0.0020	0.5077	0.4157
1.6	0.0020	0.5150	0.4410	0.0001	0.1268	0.5150	0.0020	0.5213	0.4668	0.0001	0.1083	0.5213	0.0020	0.5204	0.4671
1.7	0.0020	0.5258	0.4931	0.0000	0.0902	0.5258	0.0020	0.5303	0.5194	-0.0001	0.0720	0.5303	0.0020	0.5294	0.5196
1.8	0.0020	0.5330	0.5461	-0.0001	0.0544	0.5330	0.0020	0.5358	0.5727	-0.0002	0.0367	0.5358	0.0020	0.5349	0.5729
1.9	0.0020	0.5367	0.5996	-0.0002	0.0197	0.5367	0.0020	0.5377	0.6264	-0.0003	0.0026	0.5377	0.0020	0.5369	0.6265
2.0	0.0020	0.5370	0.6533	-0.0004	-0.0137	0.5370	0.0020	0.5363	0.6802	-0.0004	-0.0301	0.5363	0.0020	0.5355	0.6801

(b) Right-Hand Side Table

Eqs.	(6.48)			(6.49)			(6.50)				(6.51)		(6.52)	
T	$\left.\frac{dg}{dt}\right|_{0-\frac{1}{2},83}$	$\left.\frac{d\omega_I}{dt}\right|_{0-\frac{1}{2},83}$	$\left.\frac{d\theta_I}{dt}\right|_{0-\frac{1}{2},83}$	$i(t_0+h)$	$\omega_I(t_0+h)$	$\theta_I(t_0+h)$	$\left.\frac{dg}{dt}\right|_{0+h}$	$\left.\frac{d\omega_I}{dt}\right|_{0+h}$	$\left.\frac{d\theta_I}{dt}\right|_{0+h}$	$i(t_0+h)$	$\omega_I(t_0+h)$	$\theta_I(t_0+h)$	Analytical	Error × 1000
0.0	0.0023	0.0998	0.0065	0.0002	0.0100	0.0006	0.0023	0.2117	0.0100	0.0002	0.0112	0.0004	0.0000	-0.0156
0.1	0.0022	0.2676	0.0255	0.0005	0.0379	0.0029	0.0021	0.3364	0.0379	0.0005	0.0387	0.0028	0.0004	-0.0158
0.2	0.0021	0.3679	0.0578	0.0007	0.0755	0.0085	0.0020	0.4079	0.0755	0.0007	0.0761	0.0084	0.0028	-0.0167
0.3	0.0019	0.4229	0.0976	0.0008	0.1183	0.0182	0.0018	0.4435	0.1183	0.0008	0.1187	0.0182	0.0085	-0.0226
0.4	0.0018	0.4474	0.1414	0.0010	0.1634	0.0323	0.0017	0.4551	0.1634	0.0010	0.1637	0.0323	0.0182	-0.0378
0.5	0.0016	0.4514	0.1864	0.0012	0.2088	0.0509	0.0016	0.4502	0.2088	0.0012	0.2090	0.0509	0.0323	-0.0657
0.6	0.0015	0.4413	0.2312	0.0013	0.2531	0.0740	0.0014	0.4341	0.2531	0.0013	0.2532	0.0740	0.0510	-0.1091
0.7	0.0013	0.4216	0.2744	0.0015	0.2954	0.1015	0.0013	0.4103	0.2954	0.0015	0.2955	0.1015	0.0742	-0.1695
0.8	0.0012	0.3954	0.3153	0.0016	0.3350	0.1330	0.0011	0.3813	0.3350	0.0016	0.3350	0.1330	0.1017	-0.2479
0.9	0.0011	0.3647	0.3533	0.0017	0.3715	0.1684	0.0010	0.3487	0.3715	0.0017	0.3715	0.1684	0.1334	-0.3447
1.0	0.0009	0.3310	0.3881	0.0018	0.4046	0.2072	0.0008	0.3136	0.4046	0.0018	0.4047	0.2072	0.1688	-0.4596
1.1	0.0008	0.2952	0.4194	0.0019	0.4342	0.2492	0.0007	0.2771	0.4342	0.0019	0.4342	0.2492	0.2078	-0.5920
1.2	0.0006	0.2583	0.4471	0.0019	0.4600	0.2939	0.0005	0.2397	0.4600	0.0019	0.4600	0.2939	0.2499	-0.7410
1.3	0.0005	0.2207	0.4711	0.0020	0.4821	0.3410	0.0004	0.2019	0.4821	0.0020	0.4821	0.3411	0.2948	-0.9054
1.4	0.0003	0.1830	0.4913	0.0020	0.5004	0.3902	0.0003	0.1642	0.5004	0.0020	0.5004	0.3902	0.3422	-1.0840
1.5	0.0002	0.1455	0.5077	0.0020	0.5150	0.4410	0.0001	0.1269	0.5150	0.0020	0.5150	0.4410	0.3915	-1.2751
1.6	0.0001	0.1085	0.5204	0.0020	0.5258	0.4931	0.0000	0.0902	0.5258	0.0020	0.5258	0.4931	0.4425	-1.4774
1.7	-0.0001	0.0722	0.5294	0.0020	0.5330	0.5460	-0.0001	0.0544	0.5330	0.0020	0.5330	0.5461	0.4948	-1.6892
1.8	-0.0002	0.0370	0.5349	0.0020	0.5367	0.5996	-0.0002	0.0197	0.5367	0.0020	0.5367	0.5996	0.5480	-1.9089
1.9	-0.0003	0.0030	0.5369	0.0020	0.5370	0.6533	-0.0004	-0.0137	0.5370	0.0020	0.5370	0.6533	0.6017	-2.1350
2.0	-0.0004	-0.0297	0.5355	0.0019	0.5340	0.7069	-0.0005	-0.0457	0.5340	0.0019	0.5340	0.7069	0.6557	-2.3657

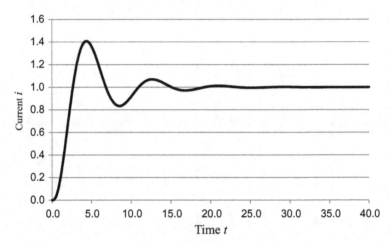

FIGURE 6.8   Plotted data from Table 6.3 (extended to $t = 40.0$).

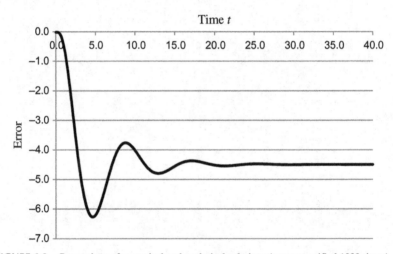

FIGURE 6.9   Comparison of numerical and analytical solutions (error magnified 1000 times).

Chapter 7

Industrial Systems

7.1 INDUSTRIAL SYSTEMS

The application of mathematical modeling for the purpose of increasing productivity is a complex system optimization exercise in any enterprise. Many are concerned about how the technology could cope with the system requirements of each implementation.

The concept of modeling manufacturing systems with more mathematical approach has connection that can be traced back to the 1960s. Using a mathematical representation of the industry system, it is expected that behavior of the system can be predicted beforehand, so that modifications can be made more effectively and less costly. Having a predicted system behavior prior to actual operation of the system can also help reduce the possibility of errors in the change process. More recently, the concept of agile manufacturing is introduced. The system characteristic, i.e., agility, which means highly adaptable to different system conditions dynamically, demands more frequent use of industry system models to examine the effect of frequent system changes so as to manage these changes effectively.

This chapter examines some typical industrial system models and discusses the methods of finding the numerical solutions for these problems.

7.2 TRANSPORT SYSTEMS MODELING

Physical transport activities in warehouses as well as supply chains are always a concern among business managers, operators, and owners. Customer satisfaction in distribution supply chains depends on fast reliable means of moving goods. Moving of goods within the warehouse or on the road does not gain any value to the distribution business and should be minimized. This section models both cases and applies some numerical methods to find an optimum operating setting.

7.2.1 Transport on a Plane (Two-Dimensional Space)

Normally, a supply chain has a number of warehouses that hold goods for temporary storage and allow retailer order picking and packing.

Demystifying Numerical Models. DOI: https://doi.org/10.1016/B978-0-08-100975-8.00007-2

The decision for selecting stock keeping units location in a warehouse is a common problem in supply chains. However, explicit consideration of both qualitative and quantitative factors during development of an optimum solution is difficult. Qualitative factors are not easily measured on a performance scale or numerical criteria. Consequently, it is difficult to combine qualitative and quantitative factors in quantitative manner such that a satisfactory evaluation of alternative solutions is obtained. In the following development, quantitative objectives are optimized. Once the quantitative objectives are evaluated, qualitative factors could be considered with other methods such as analytic hierarchy process or multicriteria decision-making process.

Similarly, the problem of determining the location of a new facility in relation to a number of existing facilities or storage is basically a transport cost minimization problem. The essence is to develop a mathematical model that represents, without loss of generosity, the total cost of transportation in the new environment. Some typical examples are:

- A new machine in a large manufacturing workshop
- A new factory in a city where there are other related factories
- A new warehouse supplying a number of production facilities and customers
- A new hospital, fire station, or library in a metropolitan area
- A new classroom building on an existing campus
- A new maintenance service depot for a bus company
- A new loading dock for a warehouse
- A new power generating plant in an urban area
- A new copying machine located in a library

A general formulation of the problem can be defined as follows:

- Existing m facilities are located at known distinct points denoted by $\mathbf{P}_1$, ..., $\mathbf{P}_m$.
- The new facility is located at point $\mathbf{X}$, which is to be determined.
- The cost of transportation is directly proportional to the distance between the new facility and existing facilities i.
- w_i is the number of trips expected between point $\mathbf{X}$ and facility i at $\mathbf{P}_i$ in a period (e.g., a year).
- The cost of transportation per trip per unit distance is a constant k.

If d_i represents the distance traveled per trip between point $\mathbf{X}$ and facility i at $\mathbf{P}_i$, then

$$f(X) = f(x, y) = k \sum_{i=1}^{m} w_i d_i \tag{7.1}$$

The problem can then be reduced to minimizing $f(X)$ by changing the location of $\mathbf{X}$, which is characterized by the coordinates x and y which determines distance d_i.

There are many ways to calculate the distance d_i. The simplest form is the straight line distance, also known as the Euclidean distance. Assuming that the facilities are built on a two-dimensional space, i.e., multistorey buildings or fly over transportation are not considered here, and if the coordinates are represented by:

$$\mathbf{P}_i = (a_i, b_i)$$

$$\mathbf{X} = (x, y)$$

the distance between the points is given by:

$$d_i = \sqrt{(x - a_i)^2 + (y - b_i)^2} \tag{7.2}$$

This means that the goal is to find the local minimum where the cost per unit distance traveled is a constant k.

$$f(X) = \sum_{i=1}^{m} kw_i \sqrt{(x - a_i)^2 + (y - b_i)^2} \tag{7.3}$$

We use a numerical example to illustrate the process of finding the solution. Assuming that there are four points in the plane denoted by $\mathbf{P}_i = \{(4,2), (8,5), (11,8), (13,2)\}$. A location $\mathbf{X}$ is required to serve the demands at these locations. The frequency of trips in a day from the desirable location $\mathbf{X}$ to $\mathbf{P}_i$ is given by $w_i = \{1,2,2,1\}$. To simplify the illustration, let $k = 1$. Substituting $\mathbf{P}_i$ into Eq. (7.3), we get

$$f(X) = \sqrt{(x - 4)^2 + (y - 2)^2} + 2\sqrt{(x - 8)^2 + (y - 5)^2} + 2\sqrt{(x - 11)^2 + (y - 8)^2}$$
$$+ \sqrt{(x - 13)^2 + (y - 2)^2}$$

$$\tag{7.4}$$

It is noted that $\mathbf{X}$ is actually one of the points in $\mathbf{P}_i$. Analytical method by differentiating Eq. (7.3) partially by x and y and equate to zero will give

$$\frac{\partial f(X)}{\partial x} = \sum_{i=1}^{m} kw_i \frac{x - a_i}{\sqrt{(x - a_i)^2 + (y - b_i)^2}} = 0 \tag{7.5}$$

$$\frac{\partial f(X)}{\partial y} = \sum_{i=1}^{m} kw_i \frac{y - b_i}{\sqrt{(x - a_i)^2 + (y - b_i)^2}} = 0 \tag{7.6}$$

The desirable $\mathbf{X} = (8, 5)$ is a singular solution resulting in unpredictable outcomes.

Graphically, Eq. (7.4) can be visualized as shown in Fig. 7.1.

A unique local minimum exists somewhere in the range of x (horizontally) and y (pointing into picture) in Fig. 7.1. To find the minimum point, Table 7.1 is set up with x and y as variables and the minimum value is found

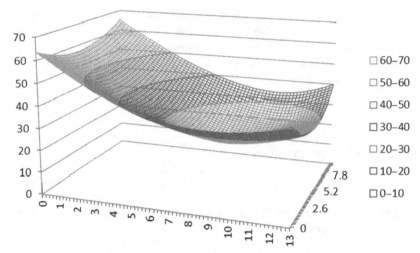

FIGURE 7.1 Surface representing the cost function.

by direct search method. Values from $(x, y) = (0,0)$ to $(13,10)$ are evaluated. The minimum value is found at $(x, y) = (8, 5)$. Numerically, this is shown by the intersection of the line of values marked in bolt font drawn from $x = 8$ and $y = 5$ as shown in Table 7.1. The series shows that for $x = 8$, the local value changes from 36.9 (when $y = 0$) to 19.3 (when $y = 5$), then 35.6 (when $y = 10$).

7.2.2 Transport in Rectilinear Layout

In practice, the passage between two points in facilities such as warehouses is not a direct line of sight. Travel occurs along a set of aisles arranged in a rectangular pattern parallel to the walls of the building or storage racks. The distance travelled is referred as rectilinear or metropolitan, and can be represented by:

$$d_i = |x - a_i| + |y - b_i| \tag{7.7}$$

Eq. (7.7) is also applicable to urban location analyses where travel occurs along an orthogonal city street design, or offices that employ a rectilinear aisle and hallway layout.

Irrespective of the form of distance d_i, the total cost defined in Eq. (7.1) needs to be minimized. Interestingly, the rectilinear distance formula combines the property of being an appropriate distance measure for a large number of location problems but it cannot be solved using analytical solution. As shown in Fig. 7.2, there are infinite paths between the points **X** and **P$_i$** all of the same length. This is not the case with Euclidean distance.

TABLE 7.1 Direct Search by Laying Out All x and y Values

x		0	1	2	3	4	5	6	7	8	9	10	Minimum
	0	63.7	61.1	59.1	57.8	57.2	57.1	57.7	58.9	60.6	62.8	65.4	57.07512
	1	58.6	55.7	53.6	52.1	51.5	51.5	52.2	53.5	55.4	57.7	60.6	51.38738
	2	53.7	50.5	48	46.5	45.9	46.0	46.8	48.2	50.3	52.9	55.9	45.82932
	3	49.2	45.5	42.7	41.1	40.5	40.7	41.6	43.2	45.4	48.2	51.5	40.49621
	4	45.3	41.2	37.4	36.2	35.6	35.7	36.7	38.4	40.8	43.9	47.4	35.54694
	5	42.1	37.9	34.5	32.3	31.2	31.1	32.0	33.9	36.6	39.9	43.7	31.04342
	6	39.7	35.5	31.8	29.1	27.4	26.9	27.8	29.8	32.8	36.3	40.4	26.87277
	7	38.0	33.6	29.7	26.5	24.1	23.0	24.0	26.4	29.5	33.3	37.7	22.95084
	8	36.9	32.5	28.4	24.9	21.9	19.3	21.3	23.8	27.0	31.0	35.6	19.31623
	9	36.5	32.0	28.0	24.5	21.6	20.0	20.5	22.4	25.3	29.4	34.2	19.9803
	10	36.8	32.3	28.4	25.1	22.6	21.3	21.2	22.1	24.4	28.6	33.8	21.05941
	11	37.8	33.3	29.5	26.5	24.4	23.2	22.9	23.2	24.0	29.2	34.5	22.85895
	12	39.4	34.9	31.2	28.6	27.0	26.0	25.8	26.3	28.1	31.8	36.7	25.77766
	13	41.9	37.4	33.3	31.6	30.4	29.7	29.7	30.5	32.5	35.7	39.8	29.61324

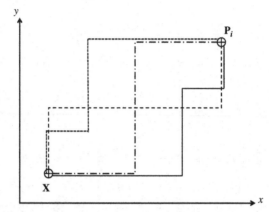

FIGURE 7.2 Different rectilinear paths between two points having same lengths.

The rectilinear distance transport problem can be stated mathematically as:

$$\min f(x,y) = \min \left[k \sum_{i=1}^{m} w_i \left(|x - a_i| + |y - b_i| \right) \right]$$ (7.8)

Eq. (7.8) can be equivalently stated as:

$$\min f(x,y) = \min \left(k \sum_{i=1}^{m} w_i |x - a_i| \right) + \min \left(k \sum_{i=1}^{m} w_i |y - b_i| \right)$$ (7.9)

Since the coordinates are orthogonal, Eq. (7.9) can be treated as separate optimization problems:

$$\begin{aligned} \min f(x,y) \Big|_x &= \min \left(k \sum_{i=1}^{m} w_i |x - a_i| \right) \\ \min f(x,y) \Big|_y &= \min \left(k \sum_{i=1}^{m} w_i |y - b_i| \right) \end{aligned}$$ (7.10)

Since this problem cannot be solved analytically, the concept for solving this problem numerically is illustrated in Fig. 7.3. In Fig. 7.2, an approach using the concept of contour lines of the cost function is used. In two-dimensional sense, a contour line is a line of constant cost in the plane. Locating the new facility at any point on a given contour line results in the same total cost. Contour lines provide considerable insight into the change of cost function $f(X)$ on the surface.

The principle is to find the local minimum when one of the variables is kept constant. This process is repeated for both x and y directions until the change of x and y coordinate of the target position is within a limit. In Fig. 7.3, the first line, vertical line (1) is drawn with a constant x value. The

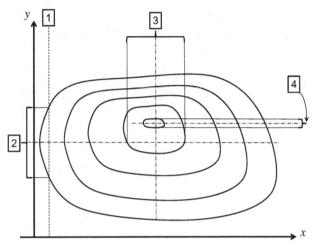

FIGURE 7.3 Finding the local minimum using contour search.

section that crosses the contour represents the range of y values that contain possible local minimal. The second line, horizontal line (2) is then constructed at the mid-point of the y range. The line cuts several contours. Choose the section that crosses the minimum contour. The third line, vertical line (3) is then constructed at the mid-point of the x range of that section. Line (3) cuts several contours and the section with the lowest contour value is selected. Similarly, the fourth line, horizontal line (4) is constructed at the mid-point of the y range. This process is repeated until the desirable accuracy of the x and y coordinates is reached. This process is illustrated with a simple example.

We use the same example to illustrate the process, i.e., there are four points in the plane denoted by $P_i = \{(4,2), (8,5), (11,8), (13,2)\}$ with $k = 1$. The frequency of trips in a day from location $\mathbf{X}$ to P_i is $w_i = \{1,2,2,1\}$.

We start the process by constructing a vertical line at $x = 0$ cutting through the contour as presented in Table 7.2 with an increment of 0.2. f is calculated according to Eq. (7.10). The (x,f) pairs marked in bold font are the pairs with minimum f values. This notation applies to Tables 7.3−7.13.

The vertical line cuts a range $(8.0 \leq y \leq 11.0)$ with minimum $f = 45.0$ as highlighted. The second horizontal line at $y = (8.0 + 11.0)/2 = 9.5$ is then constructed as presented in Table 7.3.

The horizontal line cuts a point $y = 5.0$ with minimum $f = 27.0$. The third line at $x = 5.0$ is then constructed in Table 7.4.

Table 7.4 presents that a range $(8.0 \leq y \leq 11.0)$ gives minimum $f = 27.0$. Repeat the same process with y and then x in Tables 7.5−7.7.

Tables 7.6 and 7.7 present that the optimum location for $\mathbf{X}$ is not a unique location. It lies on a line with $x = 5.00$ in a range $(8.00 \leq y \leq 11.00)$. This

TABLE 7.2 First Cut Line x = 0.0, Increment = 0.2

y	7.0	7.2	7.4	7.6	7.8	8.0	8.2	8.4	8.6	8.8	9.0	9.2	9.4
f	49.0	48.2	47.4	46.6	45.8	45.0	45.0	45.0	45.0	45.0	45.0	45.0	45.0
y	9.6	9.8	10.0	10.2	10.4	10.6	10.8	11.0	11.2	11.4	11.6	11.8	12
f	45.0	45.0	45.0	45.0	45.0	45.0	45.0	45.0	45.8	46.6	47.4	48.2	49.0

TABLE 7.3 Second Cut Line y = 9.5, Increment = 0.2

x	4.0	4.2	4.4	4.6	4.8	5.0	5.2	5.4	5.6	5.8	6.0	6.2	6.4
f	29.0	28.6	28.2	27.8	27.4	27.0	27.4	27.8	28.2	28.6	29.0	29.4	29.8

TABLE 7.4 Third Cut Line x = 5.0, Increment = 0.1

y	7.80	7.90	8.00	8.10	8.20	8.30	8.40	8.50	8.60	8.70
f	27.8	27.4	27.0	27.0	27.0	27.0	27.0	27.0	27.0	27.0
y	8.80	8.90	9.00	9.10	9.20	9.30	9.40	9.50	9.60	9.70
f	27.0	27.0	27.0	27.0	27.0	27.0	27.0	27.0	27.0	27.0
y	9.80	9.90	10.00	10.10	10.20	10.30	10.40	10.50	10.60	10.70
f	27.0	27.0	27.0	27.0	27.0	27.0	27.0	27.0	27.0	27.0
y	10.80	10.90	11.00	11.10	11.20					
f	27.0	27.0	27.0	27.4	27.8					

TABLE 7.5 Fourth Cut Line y = (8.0 + 11.0)/2 = 9.5, Increment = 0.1

x	4.80	4.90	**5.00**	5.10	5.20
f	27.40	27.20	**27.00**	27.20	27.40

means that the final location can be anywhere on the line $x = 5.0$, between $(x, y) = (5,8)$ and $(5,11)$. The solution is not a single point and this actually offers more flexibility to the manufacturing system engineer to work with.

7.2.3   Transport in a Building

In cities where land is precious, multistorey buildings are often used as warehouses. Some existing locations could be on a different floor. The distance needs to be taken as a three-dimensional space. An additional term is added to form Eq. (7.11).

$$d_i = |x - a_i| + \left|y - b_i\right| + |z - c_i| \tag{7.11}$$

The basic numerical solution process is to set two of the three variables constant while exploring the range of the third variable cutting the contour. We extend the previous example to the third dimension with four points denoted by $P_i = \{(4,2,5), (8,5,2), (11,8,7), (13,25)\}$. A location $\mathbf{X}$ is required to serve the demands at these locations. The frequency of trips in a day from the desirable location $\mathbf{X}$ to P_i remains unchanged $w_i = \{1,2, 2,1\}$.

Let's start with setting $x = 0$ and $y = 0$. Both variables are kept constant while exploring the distance function by changing z from 0 as presented in Table 7.8.

The minimum cut across the contour is $z = 5$. The process is repeated for this z value and changing y (Table 7.9).

The minimum cut across the contour occurs at $y = 5$. The process is repeated for this y value while keeping previous z value and changing x (Table 7.10).

The minimum cut across the contour occurs at $(8.0 \le x \le 11.0)$. The minimum distance value on this contour is constant. Next, the process is repeated for the mid-point of this range, i.e., $x = 9.5$ with $y = 5$ while changing z values with a smaller increment. To demonstrate reaching the answer quickly, an increment of 0.01 is used in Table 7.11.

The minimum contour section occurs at $(4.98 \le z \le 5.02)$. The mid-point of this range is $z = 5.00$. The same process is repeated for changing y and then x in Tables 7.12 and 7.13.

Tables 7.11−7.13 present that the area bounded by { $(7.99 \le x \le 11.01)$, $(4.98 \le y \le 5.02)$, $(4.98 \le z \le 5.02)$ } has the shortest absolute distance from any of the four destinations. The location $\mathbf{X}$ can be anywhere within these boundaries.

TABLE 7.6 Fifth Cut Line x = 5.0, Increment = 0.01 (Values Between x = 8.11 and x = 10.94 Are Omitted)

y	7.90	7.91	7.92	7.93	7.94	7.95	7.96	7.97	7.98	7.99
F	27.40	27.36	27.32	27.28	27.24	27.20	27.16	27.12	27.08	27.04
y	8.00	8.01	8.02	8.03	8.04	8.05	8.06	8.07	8.08	8.09
F	27.00	27.00	27.00	27.00	27.00	27.00	27.00	27.00	27.00	27.00
y	8.10	...	...	...	10.95	10.96	10.97	10.98	10.99	11.00
f	27.00	...	...	...	27.00	27.00	27.00	27.00	27.00	27.00
y	11.01	11.02	11.03	11.04	11.05	11.06	11.07	11.08	11.09	11.10
f	27.04	27.08	27.12	27.16	27.20	27.24	27.28	27.32	27.36	27.40

TABLE 7.7 Sixth Cut Line y = (8.00 + 11.00)/2 = 9.50, Increment = 0.01

x	4.96	4.97	4.98	4.99	5.00	5.01	5.02	5.03	5.04
f	27.08	27.06	27.04	27.02	27.00	27.02	27.04	27.06	27.08

TABLE 7.8 First Cut Line x = 0.0, y = 0.0, Increment = 0.2

z	0.0	0.2	0.4	0.6	0.8	1.0	1.2	1.4	1.6	1.8	2.0	2.2
f	113.0	111.8	110.6	109.4	108.2	107.0	105.8	104.6	103.4	102.2	101.0	100.6
z	2.4	2.6	2.8	3.0	3.2	3.4	3.6	3.8	4.0	4.2	4.4	4.6
f	100.2	99.8	99.4	99.0	98.6	98.2	97.8	97.4	97.0	96.6	96.2	95.8
z	4.8	5.0	5.2	5.4	5.6	5.8	6.0	6.2	6.4	6.6	6.8	7.0
f	95.4	95.0	95.4	95.8	96.2	96.6	97.0	97.4	97.8	98.2	98.6	99.0

TABLE 7.9 Second Cut Line x = 0.0, z = 5.0, Increment = 0.2

	0.0	0.2	0.4	0.6	0.8	1	1.2	1.4	1.6	1.8	2	2.2
y	0.0	0.2	0.4	0.6	0.8	1	1.2	1.4	1.6	1.8	2	2.2
f	95.0	93.8	92.6	91.4	90.2	89.0	87.8	86.6	85.4	84.2	83.0	82.6
y	2.4	2.6	2.8	3	3.2	3.4	3.6	3.8	4	4.2	4.4	4.6
f	82.2	81.8	81.4	81.0	80.6	80.2	79.8	79.4	79.0	78.6	78.2	77.8
y	4.8	**5**	5.2	5.4	5.6	5.8	6	6.2	6.4	6.6	6.8	7
f	77.4	**77.0**	77.4	77.8	78.2	78.6	79.0	79.4	79.8	80.2	80.6	81.0

TABLE 7.10 Third Cut Line y = 5.0, z = 5.0, Increment = 0.2

	0.0	0.2	0.4	0.6	0.8	1	1.2	1.4	1.6	1.8	2	2.2	2.4	2.6	2.8	3
x	0.0	0.2	0.4	0.6	0.8	1	1.2	1.4	1.6	1.8	2	2.2	2.4	2.6	2.8	3
f	77.0	75.8	74.6	73.4	72.2	71.0	69.8	68.6	67.4	66.2	65.0	63.8	62.6	61.4	60.2	59.0
x	3.2	3.4	3.6	3.8	4	4.2	4.4	4.6	4.8	5	5.2	5.4	5.6	5.8	6	6.2
f	57.8	56.6	55.4	54.2	53.0	52.2	51.4	50.6	49.8	49.0	48.2	47.4	46.6	45.8	45.0	44.2
x	6.4	6.6	6.8	7	7.2	7.4	7.6	7.8	**8**	**8.2**	**8.4**	**8.6**	**8.8**	**9**	**9.2**	**9.4**
f	43.4	42.6	41.8	41.0	40.2	39.4	38.6	37.8	**37.0**	**37.0**	**37.0**	**37.0**	**37.0**	**37.0**	**37.0**	**37.0**
x	**9.6**	**9.8**	**10**	**10.2**	**10.4**	**10.6**	**10.8**	**11**	11.2	11.4	11.6	11.8	12	12.2	12.4	12.6
f	**37.0**	**37.0**	**37.0**	**37.0**	**37.0**	**37.0**	**37.0**	**37.0**	37.8	38.6	39.4	40.2	41.0	41.8	42.6	43.4

TABLE 7.11 Fourth Cut Line x = 9.5, y = 5.0, Increment = 0.01

z	4.8	4.81	4.82	4.83	4.84	4.85	4.86	4.87	4.88	4.89	4.9	4.91	4.92
f	37.4	37.4	37.4	37.3	37.3	37.3	37.3	37.3	37.2	37.2	37.2	37.2	37.2
z	4.93	4.94	4.95	4.96	4.97	4.98	4.99	5.00	5.01	5.02	5.03	5.04	5.05
f	37.1	37.1	37.1	37.1	37.1	37.0	37.0	37.0	37.0	37.0	37.1	37.1	37.1

TABLE 7.12 Fifth Cut Line x = 9.5, z = 5.0, Increment = 0.01

y	4.8	4.81	4.82	4.83	4.84	4.85	4.86	4.87	4.88	4.89	4.9	4.91	4.92
f	37.4	37.4	37.4	37.3	37.3	37.3	37.3	37.3	37.2	37.2	37.2	37.2	37.2
y	4.93	4.94	4.95	4.96	4.97	4.98	4.99	5.00	5.01	5.02	5.03	5.04	5.05
f	37.1	37.1	37.1	37.1	37.1	37.0	37.0	37.0	37.0	37.0	37.1	37.1	37.1

TABLE 7.13 Sixth Cut Line y = 5.0, z = 5.0, Increment = 0.01

x	7.8	7.81	7.82	7.83	7.84	7.85	7.86	7.87	7.88	7.89	7.9	7.91	7.92
f	37.8	37.8	37.7	37.7	37.6	37.6	37.6	37.5	37.5	37.4	37.4	37.4	37.3
x	7.93	7.94	7.95	7.96	7.97	7.98	7.99	8	8.01	8.02	8.03	8.04	8.05
f	37.3	37.2	37.2	37.1	37.1	37.1	37.0	37.0	37.0	37.0	37.0	37.0	37.0
x	8.06	8.07	8.08	8.09	8.10	...	...	...	...	10.94	10.95	10.96	10.97
f	37.0	37.0	37.0	37.0	37.0	...	...	...	...	37.0	37.0	37.0	37.0
x	10.98	10.99	11	11.01	11.02	11.03	11.04	11.05	11.06	11.07	11.08	11.09	11.1
f	37.0	37.0	37.0	37.0	37.1	37.1	37.2	37.2	37.2	37.3	37.3	37.4	37.4

7.3 INVENTORY CONTROL

Inventory can be regarded as a form of temporary storage of materials, machines, money, or data for an organization at some point in time. Inventory changes over time. The changes are caused by one or more other functions in the organization. If the changes are stopped for some reason, e.g., end of year stock-take, the inventory is said to be taken as a snapshot at that moment. Inventory does not come automatically. It is built up as a result of use of resources. Hence, it is often conceived as an asset that is idle.

The inventory is an active account recording the movement of goods "added to" and "depleted from" the storage, while maintaining a computational process to determine the balance of items in the inventory. Due to some process errors over time, the number of items recorded in the balance of the inventory is not always the same as the actual number in the storage. Typically, an annual stock-take is an activity of the organization to ascertain the accuracy of the inventory on record compared with the inventory actually held in the storage. The annual stock-take is therefore a quality assurance exercise and is not in the scope of this chapter. The interest here is on how to represent the movement of goods in a mathematical model and how to use this model to develop the parameters to control the inventory for the benefit of the industrial operation.

7.3.1 Replenish Directly From a Stock Pile

The simplest inventory model is the so-called "economic order quantity" model. The model describes how the level of inventory varies in time under the following conditions:

- The demand (sales) per unit time is constant, represented by D.
- The quantity replenished when the inventory runs out is Q.
- Replenishment comes from an external stock pile that can supply to this inventory immediately (no lead time).

When this model is associated with two cost items:

- Cost of holding one unit per unit time, C_h
- Cost of ordering per transaction, C_o

The economic order quantity is given by:

$$Q^* = \sqrt{\frac{2DC_o}{C_h}} \tag{7.12}$$

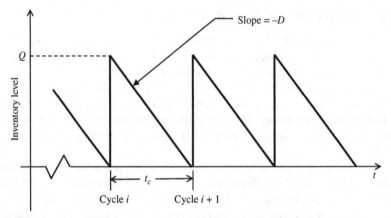

FIGURE 7.4   Simple inventory model—replenish from a stock pile.

The inventory level is then represented by a simple saw tooth graph as shown in Fig. 7.4, where t_c is the inventory cycle time, and i is the inventory cycle number.

$$s(t) = Q - D(t - it_c) \quad \text{for } t \in [it_c, (i + 1)t_c], \quad i = 0, 1, 2, \ldots, n \quad (7.13)$$

Now consider a more practical situation. If the delivery time of the replenish quantity takes one day, i.e., order for replenishment is raised today, due to document flow requirements, the quantity is delivered the next day from the stock pile. Let's redefine the terms as follows:

$q(i) =$ replenish quantity in day i. Note that i is now a discrete day number.

$s_o(i) =$ quantity stock-out, i.e., number of items owing the customer

$D =$ demand in per day

The problem can then be re-written as an inventory model in Eq. (7.13):

$$s(i) = \begin{cases} s(i-1) + q(i-1) - D + s_o(i-1) \\ 0 \end{cases} \text{if} \quad \begin{array}{l} s(i-1) + q(i-1) - D + s_o(i-1) \geq 0 \\ \\ s(i-1) + q(i-1) - D + s_o(i-1) < 0 \end{array}$$

$$(7.14)$$

for

$$i \in [1, n]$$

The re-order point is defined as a trigger by the inventory level falling below certain value:

$$q(i) = \begin{cases} Q \\ 0 \end{cases} \text{if} \quad \begin{array}{l} s(i) \leq Q_R \\ s(i) > Q_R \end{array} \quad (7.15)$$

where Q_R is the quantity triggering an order to replenish, i.e., the re-order point. The stock-out quantity can be computed as:

$$s_o(i) = \begin{cases} s(i-1)+q(i-1)-D+s_o(i-1) & \text{if } s(i-1)+q(i-1)-D+s_o(i-1)<0 \\ 0 & s(i-1)+q(i-1)-D+s_o(i-1)\geq0 \end{cases}$$

(7.16)

Note that $s_o(i)$ is negative in Eq. (7.16). A computational table can then be created to examine the effect of setting different values of Q, Q_R. To illustrate the model and numerical solution to the model, a simple example is used.

Assume that $Q = 38$ and $D = 33$, both are constants. It is also assumed that the initial stock level $s(0) = 38$ and $Q_R = 50$. Table 7.14 presents how

TABLE 7.14 Numerical Solution of Simple Inventory Model for $Q_R = 50$

Day i	D	$s(i)$	$s_o(i)$	$q(i)$
0	0	38	0	0
1		5	0	38
2		10	0	38
3		15	0	38
4		20	0	38
5		25	0	38
6		30	0	38
7		35	0	38
8		40	0	38
9		45	0	38
10	33	50	0	38
11		55	0	0
12		22	0	38
13		27	0	38
14		32	0	38
15		37	0	38
16		42	0	38
17		47	0	38
18		52	0	0
19		19	0	38
20		24	0	38

the computation table can be set up for computing the stock level, stock-out, and ordered quantity in different period $i = 0$ to 20. Note that there is no demand on day 0 because no order is received prior to start of this period, viz, the system is assumed not operational prior to day 0. The initial stock level $s(0) = 38$ as given and there is no stock-out $s_o(0) = 0$, and no replenished goods $q(0) = 0$.

Referring to Table 7.14, on day $i = 1$, from Eq. (7.14),

$$s(1) = \begin{cases} s(0) + q(0) - D + s_o(0) = 38 - 0 - 33 + 0 = 5 \geq 0 \\ 0 \end{cases} \Rightarrow s(1) = 5$$

(7.17)

The re-order point is computed from Eq. (7.15),

$$q(1) = \begin{cases} 38 \\ 0 \end{cases} \text{ if } \begin{array}{l} s(1) \leq 50 \\ s(1) > 50 \end{array} \text{ which gives } q(1) = 38 \qquad (7.18)$$

The stock-out value is computed from Eq. (7.16),

$$s_o(1) = \begin{cases} s(0) + q(0) - D + s_o(0) \\ 0 \end{cases} \text{ if } \begin{array}{l} s(0) + q(0) - D + s_o(0) < 0 \\ s(0) + q(0) - D + s_o(0) \geq 0 \end{array} \qquad (7.19)$$

This gives,

$$s_o(1) = 0 \qquad (7.20)$$

The other rows are computed similarly.

It can be seen from Table 7.14 that there is no stock-out situation. The average stock can be found by averaging all values of $s(i)$, which gives 31.905. This means that we are keeping almost the same level as the daily demand in our store, which is not a desirable situation. High inventory requires more space (more rent paid), more people to handle (more labor cost), more chance of damage (rejects by customer), etc.

To investigate the optimal stock level, Table 7.15 is a repeat of the logic in Table 7.13 at different Q_R levels.

The average stock-out can be plotted against Q_R as shown in Fig. 7.5. Note that stock-out is represented by a negative number. It can be seen from Table 7.14 that if the re-order point is set too low, stock-out will occur. The highest loss occurs at $Q_R = 0$, i.e., re-order only when there is no stock. From the sales' perspective, any stock-out situation is not acceptable. Lost sale may occur, but if management is prepared to take the risk so as to save cost, then reducing inventory could be a viable choice.

Fig. 7.5 shows that stock-out reduces from the highest loss of 14 units to zero as the re-order point increases. The corresponding stock level also increases even after the average stock-out levels out at zero. It is clear that the optimum re-order point is where the average stock-out just turns zero.

TABLE 7.15 Effect of Re-order Point to Stock-Out in Simple Inventory Model With Constant Demand

$Q_R =$	0	10	20	30	40	50	60
Day i							
0	0	0	0	0	0	0	0
1	0	0	0	0	0	0	0
2	−28	0	0	0	0	0	0
3	−23	0	0	0	0	0	0
4	−18	−18	0	0	0	0	0
5	−13	−13	0	0	0	0	0
6	−8	−8	−8	0	0	0	0
7	−3	−3	−3	0	0	0	0
8	0	0	0	0	0	0	0
9	−31	0	0	0	0	0	0
10	−26	0	0	0	0	0	0
11	−21	−21	0	0	0	0	0
12	−16	−16	0	0	0	0	0
13	−11	−11	−11	0	0	0	0
14	−6	−6	−6	0	0	0	0
15	−1	−1	−1	−1	0	0	0
16	0	0	0	0	0	0	0
17	−29	0	0	0	0	0	0
18	−24	0	0	0	0	0	0
19	−19	−19	0	0	0	0	0
20	−14	−14	0	0	0	0	0
Average	−13.857	−6.190	−1.381	−0.048	0.000	0.000	0.000

The inventory model is now available. It is possible to find the optimum operating parameters by analyzing the model. The interest will be on the re-order point that is just sufficient to keep no stock out. In this case, the bisection method is used because it does not rely on the existence of its derivatives. It appears from Fig. 7.5 that the first zero stock-out point is somewhere between $Q_R = 20$ and $Q_R = 40$. The tolerance is set at 0.01. Table 7.16 is then set up to find the first zero point.

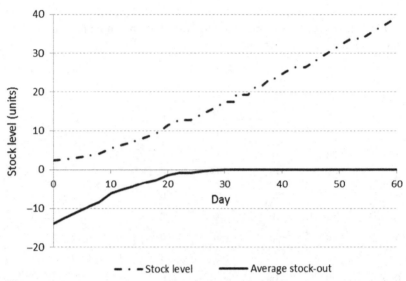

FIGURE 7.5 Effect of changing re-order point Q_R on average stock-out and stock level.

Referring to Table 7.15, in iteration 1, the starting Q_{Rs} and the ending Q_{Re} are set at 20 and 40, respectively. The mid-point Q_{Rm} is therefore 30. By logic presented in Table 7.13 after substituting these values into the equations, values of stock-out s_o are found for each of the re-order point Q_{Rs}, Q_{Re}, and Q_{Rm}. The bisection case is determined as case 1, which means that the zero point should lie between 30 and 40. The process is then repeated for other iterations. The optimum re-order point is found to be $Q_R = 32.002$, when the error from the last Q_{Rm} is $0.0098 < 0.01$.

7.3.2 Varying Demand

In reality, demand will never be constant. This can be modeled by introducing the demand function $d(i)$ which has varying values in different periods. Eq. (7.14) is then modified:

$$s(i) = \begin{cases} s(i-1)+q(i-1)-d(i)+s_o(i-1) \\ 0 \end{cases} \text{ if } \begin{array}{l} s(i-1)+q(i-1)-d(i)+s_o(i-1) \geq 0 \\ s(i-1)+q(i-1)-d(i)+s_o(i-1) < 0 \end{array}$$

(7.21)

for

$$i \in [1, n]$$

For practical reason, the re-order point is usually managed at a constant level:

$$q(i) = \begin{cases} Q \\ 0 \end{cases} \text{ if } \begin{array}{l} s(i) \leq Q_R \\ s(i) > Q_R \end{array}$$

(7.22)

TABLE 7.16 Bisection Method to Find the Optimum Re-order Point

Iteration	Q_{Rs}	Q_{Re}	Q_{Rm}	$s_o(Q_{Rs})$	$s_o(Q_{Re})$	$s_o(Q_{Rm})$	Case	ε_d
1	20.0000	40.0000	30.0000	−1.3810	0.0000	−0.0476	1	NA
2	30.0000	40.0000	35.0000	−0.0476	0.0000	0.0000	2	5.0000
3	30.0000	35.0000	32.5000	−0.0476	0.0000	0.0000	2	2.5000
4	30.0000	32.5000	31.2500	−0.0476	0.0000	−0.0476	1	1.2500
5	31.2500	32.5000	31.8750	−0.0476	0.0000	−0.0476	1	0.6250
6	31.8750	32.5000	32.1875	−0.0476	0.0000	0.0000	2	0.3125
7	31.8750	32.1875	32.0313	−0.0476	0.0000	0.0000	2	0.1563
8	31.8750	32.0313	31.9531	−0.0476	0.0000	−0.0476	1	0.0781
9	31.9531	32.0313	31.9922	−0.0476	0.0000	−0.0476	1	0.0391
10	31.9922	32.0313	32.0117	−0.0476	0.0000	0.0000	2	0.0195
11	31.9922	32.0117	32.0020	−0.0476	0.0000	0.0000	2	0.0098

The stock-out quantity can be computed similar to Eq. (7.16) as:

$$s_o(i) = \begin{cases} s(i-1) + q(i-1) - d(i) + s_o(i-1) \\ 0 \end{cases} \text{ if } \begin{array}{l} s(i-1) + q(i-1) - d(i) + s_o(i-1) < 0 \\ s(i-1) + q(i-1) - d(i) + s_o(i-1) \geq 0 \end{array}$$

(7.23)

Assume that the demand varies as presented in Table 7.17. The re-order quantity remains unchanged, i.e., $Q = 38$, the stock level, stock-out, and re-order quantity for $Q_R = 10$ is computed.

TABLE 7.17 Effect of Re-order Point to Stock Level in Simple Inventory Model With Variable Demand

Day i	$d(i)$	$s(i)$	$s_o(i)$	$q(i)$
0	0	38	0	0
1	20	18	0	0
2	27	0	−9	38
3	30	0	−1	38
4	20	17	0	0
5	31	0	−14	38
6	11	13	0	0
7	27	0	−14	38
8	20	4	0	38
9	20	22	0	0
10	21	1	0	38
11	29	10	0	38
12	25	23	0	0
13	32	0	−9	38
14	21	8	0	38
15	25	21	0	0
16	26	0	−5	38
17	29	4	0	38
18	30	12	0	0
19	32	0	−20	38
20	25	0	−7	38
Average	23.86	9.10	−3.76	23.52

The development process of Table 7.17 can be explained as follows. The situation on day $i = 0$ is similar to the fixed demand model. On day $i = 1$, from Eq. (7.21),

$$s(1) = \begin{cases} s(0) + q(0) - d(1) + s_o(0) = 38 - 0 - 20 - 0 = 18 \geq 0 \\ 0 \end{cases} \Rightarrow s(1) = 18$$

(7.24)

The re-order quantity is computed from Eq. (7.22),

$$q(1) = \begin{cases} 38 \\ 0 \end{cases} \quad \text{if} \quad \begin{matrix} s(1) \leq 10 \\ s(1) > 10 \end{matrix} \quad \text{which gives } q(1) = 0 \qquad (7.25)$$

The stock-out value is computed from Eq. (7.23),

$$s_o(1) = \begin{cases} s(0) + q(0) - d(1) + s_o(0) \\ 0 \end{cases} \quad \text{if} \quad \begin{matrix} s(0) + q(0) - d(1) + s_o(0) < 0 \\ s(0) + q(0) - d(1) + s_o(0) \geq 0 \end{matrix}$$

(7.26)

This gives,

$$s_o(1) = 0 \qquad (7.27)$$

On day $i = 2$,

$$s(2) = \begin{cases} s(1) + q(1) - d(2) + s_o(1) = 18 - 0 - 27 - 0 = -9 < 0 \\ 0 \end{cases} \Rightarrow s(1) = 0$$

(7.28)

$$q(2) = \begin{cases} 38 \\ 0 \end{cases} \quad \text{if} \quad \begin{matrix} s(2) \leq 10 \\ s(2) > 10 \end{matrix} \quad \text{which gives } q(2) = 0 \qquad (7.29)$$

$$s_o(2) = \begin{cases} s(1) + q(1) - d(2) + s_o(1) \\ 0 \end{cases} \quad \text{if} \quad \begin{matrix} s(1) + q(1) - d(2) + s_o(1) < 0 \\ s(1) + q(1) - d(2) + s_o(1) \geq 0 \end{matrix}$$

(7.30)

Hence, $s_o(2) = -9$. The other rows are computed similarly.

Expanding in the same way as the constant demand model, the stock level and stock-out are plotted in Fig. 7.6.

Again, using the bisection method, the first zero stock-out point is found within the limits $Q_R = 20$ and $Q_R = 30$ as presented in Table 7.18. The tolerance is set at 0.01.

The optimum re-order point is found to be $Q_R = 29.9941$. It should be noted that this result depends on the demand distribution. However, the process illustrated in this section should apply to generalized problems.

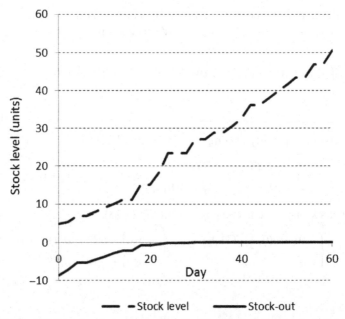

FIGURE 7.6 Effect of changing re-order point Q_R on average stock-out and stock level with varying demand.

7.3.3 Replenishing From Manufacturing

Management of inventory is not limited to a distributor situation. Manufacturing companies are often restricted by the space they have and hence controlling inventory as the product is made from the manufacturing shop floor is critical as well. The inventory model of replenishing from manufacturing describes how the level of inventory varies in time under the following conditions:

- The demand (sales) per unit time is constant, represented by D
- The quantity ordered at the re-order point Q_R is Q
- Replenishment comes from manufacturing at a constant rate Q_m per unit time such that $Q \geq Q_m$
- The manufacturing is scheduled to occur on the day of receiving the production order. Full delivery of the manufactured batch Q_m will occur the next day.

The problem can then be re-written as an inventory model as in Eq (7.14):

$$s(i) = \begin{cases} s(i-1)+q(i-1)-D+s_o(i-1) \\ 0 \end{cases} \text{ if } \begin{array}{l} s(i-1)+q(i-1)-D+s_o(i-1) \geq 0 \\ s(i-1)+q(i-1)-D+s_o(i-1) < 0 \end{array}$$

$$(7.31)$$

TABLE 7.18 Bisection Method to Find the Optimum Re-order Point

Iteration	Q_{Rs}	Q_{Re}	Q_{Rm}	$s_o(Q_{Rs})$	$s_o(Q_{Re})$	$s_o(Q_{Rm})$	Case	ε_d
1	20.0000	30.0000	25.0000	− 3.7619	− 0.0476	− 0.0476	1	NA
2	25.0000	30.0000	27.5000	− 0.0476	− 0.0476	− 0.0476	1	2.5000
3	27.5000	30.0000	28.7500	− 0.0476	− 0.0476	− 0.0476	1	1.2500
4	28.7500	30.0000	29.3750	− 0.0476	− 0.0476	0.0000	2	0.6250
5	28.7500	29.3750	29.0625	− 0.0476	0.0000	0.0000	2	0.3125
6	28.7500	29.0625	28.9063	− 0.0476	0.0000	− 0.0476	1	0.1563
7	28.9063	29.0625	28.9844	− 0.0476	0.0000	− 0.0476	1	0.0781
8	28.9844	29.0625	29.0234	− 0.0476	0.0000	0.0000	2	0.0391
9	28.9844	29.0234	29.0039	− 0.0476	0.0000	0.0000	2	0.0195
10	28.9844	29.0039	28.9941	− 0.0476	0.0000	− 0.0476	1	0.0098

for

$$i \in [1, n]$$

The re-order quantity q_{order} is triggered by the re-order point. It is assumed that the production order is kept at the pre-set order level, i.e., Q:

$$q_{order}(i) = \begin{cases} 0 \\ Q \end{cases} \text{if} \begin{cases} s(i) > Q_R \\ s(i) \leq Q_R \end{cases} \tag{7.32}$$

The replenishing quantity q comes from the manufacturing section of the organization. The quantity available at time i is governed by the manufacturing capacity:

$$q(i) = \begin{cases} Q_m & q_{order}(i) \geq Q_m \\ q_{order}(i) & \text{if} \quad q_{order}(i) < Q_m \\ 0 & q_{order}(i) = 0 \end{cases} \tag{7.33}$$

where Q_R is the quantity triggering an order to replenish, i.e., the re-order point. Similarly, the stock-out quantity can be computed as:

$$s_o(i) = \begin{cases} s(i-1) + q(i) - D + s_o(i-1) \\ 0 \end{cases} \text{if} \begin{array}{l} s(i-1) + q(i-1) - D + s_o(i-1) < 0 \\ s(i-1) + q(i-1) - D + s_o(i-1) \geq 0 \end{array} \tag{7.34}$$

Assume that $Q = 50$, $Q_m = 35$, and $D = 33$. All are constants. It is also assumed that the initial stock level $s(0) = 50$ and $Q_R = 30$. Similar to previous cases, there is no demand and no replenishment on day 0. Table 7.19 presents how the computation table can be set up for computing the stock level, stock-out, and ordered quantity in different period $i = 0$ to 20.

Referring to Table 7.19, on day $i = 1$, from Eq. (7.31),

$$s(1) = \begin{cases} s(0) + q(0) - D + s_o(0) = 50 - 0 - 33 + 0 = 17 \geq 0 \\ 0 \end{cases} \Rightarrow s(1) = 17 \tag{7.35}$$

The production order quantity is then computed from Eq. (7.32) as:

$$q_{order}(1) = \begin{cases} 0 \\ 50 \end{cases} \text{if} \begin{cases} s(1) > 30 \\ s(1) \leq 30 \end{cases} \Rightarrow q_{order}(1) = 50 \tag{7.36}$$

The manufacturing quantity on the day is computed from Eq. (7.33):

$$q(1) = \begin{cases} 35 & q_{order}(1) \geq 35 \\ 50 & \text{if} \quad q_{order}(1) < 35 \Rightarrow q(1) = 35 \\ 0 & q_{order}(1) = 0 \end{cases} \tag{7.37}$$

The stock-out value is computed from Eq. (7.34):

TABLE 7.19 Numerical Solution of Manufacturing Inventory Model for $Q_R = 30$

Day i	D	$s(i)$	$s_o(i)$	$q_{order}(i)$	$q(i)$
0	0	50	0	0	0
1		17	0	50	35
2		19	0	50	35
3		21	0	50	35
4		23	0	50	35
5		25	0	50	35
6		27	0	50	35
7		29	0	50	35
8		31	0	0	0
9		0	−2	50	35
10		0	0	50	35
11	33	2	0	50	35
12		4	0	50	35
13		6	0	50	35
14		8	0	50	35
15		10	0	50	35
16		12	0	50	35
17		14	0	50	35
18		16	0	50	35
19		18	0	50	35
20		20	0	50	35
Average		16.7619	−0.0952		

TABLE 7.20 Effect of Re-order Point to Stock-Out in Manufacturing Inventory Model With Constant Demand

$Q_R =$	0	10	20	30	40	50	60
Day i							
0	0	0	0	0	0	0	0
1	0	0	0	0	0	0	0
2	−16	−16	0	0	0	0	0
3	−14	−14	0	0	0	0	0
4	−12	−12	−12	0	0	0	0

(Continued)

TABLE 7.20 (Continued)

$Q_R =$	0	10	20	30	40	50	60
Day i							
5	−10	−10	−10	0	0	0	0
6	−8	−8	−8	0	0	0	0
7	−6	−6	−6	0	0	0	0
8	−4	−4	−4	0	0	0	0
9	−2	−2	−2	−2	0	0	0
10	0	0	0	0	0	0	0
11	0	0	0	0	0	0	0
12	−31	0	0	0	0	0	0
13	−29	0	0	0	0	0	0
14	−27	0	0	0	0	0	0
15	−25	0	0	0	0	0	0
16	−23	0	0	0	0	0	0
17	−21	−21	0	0	0	0	0
18	−19	−19	0	0	0	0	0
19	−17	−17	0	0	0	0	0
20	−15	−15	0	0	0	0	0
Av.	−13.286	−6.857	−2.000	−0.095	0.000	0.000	0.000

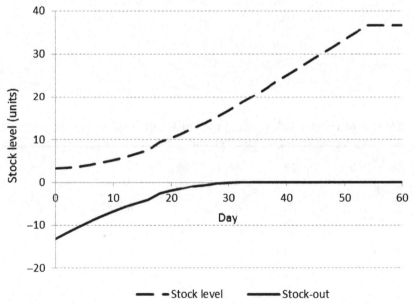

FIGURE 7.7 Effect of changing re-order point Q_R on average stock-out and stock level from manufacturing source.

TABLE 7.21 Bisection Method to Find the Optimum Re-order Point

Iteration	Q_{Rs}	Q_{Re}	Q_{Rm}	$s_o(Q_{Rs})$	$s_o(Q_{Re})$	$s_o(Q_{Rm})$	Case	ε_d
1	20.0000	40.0000	30.0000	−2.0000	0.0000	−0.0952	1	NA
2	30.0000	40.0000	35.0000	−0.0952	0.0000	0.0000	2	5.0000
3	30.0000	35.0000	32.5000	−0.0952	0.0000	0.0000	2	2.5000
4	30.0000	32.5000	31.2500	−0.0952	0.0000	0.0000	2	1.2500
5	30.0000	31.2500	30.6250	−0.0952	0.0000	−0.0952	1	0.6250
6	30.6250	31.2500	30.9375	−0.0952	0.0000	−0.0952	1	0.3125
7	30.9375	31.2500	31.0938	−0.0952	0.0000	0.0000	2	0.1563
8	30.9375	31.0938	31.0156	−0.0952	0.0000	0.0000	2	0.0781
9	30.9375	31.0156	30.9766	−0.0952	0.0000	−0.0952	1	0.0391
10	30.9766	31.0156	30.9961	−0.0952	0.0000	−0.0952	1	0.0195
11	30.9961	31.0156	31.0059	−0.0952	0.0000	0.0000	2	0.0098

$$s_o(1) = \begin{cases} s(0) + q(0) - D + s_o(0) \\ 0 \end{cases} \quad \text{if} \quad \begin{array}{l} s(0) + q(0) - D + s_o(0) < 0 \\ s(0) + q(0) - D + s_o(0) \geq 0 \end{array} \quad (7.38)$$

This gives,

$$s_o(1) = 0 \tag{7.39}$$

The other rows are computed similarly. To investigate the optimal stock level, Table 7.20 extends to other Q_R levels.

The average stock-out can be plotted against Q_R as shown in Fig. 7.7. The first stock-out, where the optimum re-order point is, can be found by bisection method as presented in Table 7.21.

The varying demand case for manufacturing inventory model can be formulated similarly and is left to the reader to work out.

Chapter 8

Systems Engineering

8.1 SYSTEMS ANALYSIS

The word system commonly refers to the cohesive functioning of a number of individually identifiable items working together. The five critical challenges discussed in Chapter 1, Introduction to Engineering Systems form the basis of analyzing systems. In a more generic sense, system analysis has to do with planning, designing, manufacturing, operating and managing to exploit functional, technical, social, environmental, and esthetic applications. It requires the creative manipulation and coordination of materials and technology, and most important of all, the stability and reliability of its performance.

The components of a system are normally represented in scientific terms as a model. Previous chapters have highlighted many of these models, some of them can be complex. A model is a representation of the real world. There is no right or wrong model. The meaning of the model as compared to that reality can be distorted. In this case the model is said to be a bad model. On the other hand, if the model produces the meaning of the real world, the model is said to be a good model. An system model combines the component models into a massive model that hopefully can be analyzed to predict its performance. It is clear from this development that as the number of components increases, the complexity of the system increases.

This chapter focuses on the representation of system model as opposed to the component models presented previously. As a system model, the priority is on the critical challenges. To analyze the impact of these challenges on the system a different perspective of modeling is required.

8.2 SYSTEM OPERATION STRATEGY

Deciding how the system runs is the most basic question to be asked in analyzing operation of engineering machines running in parallel. Before illustrating with an example, a few fundamental concepts in reliability need to be clarified.

There are many mathematical functions that can be used for estimating the reliability of a system. The most commonly used function is the

Demystifying Numerical Models. DOI: https://doi.org/10.1016/B978-0-08-100975-8.00008-4
155

exponential function, which states that the reliability of the system over time t is given by

$$R(t) = \exp(-\lambda t) \tag{8.1}$$

where λ is the failure rate, which is expressed as the number of failures over a standard time. Although the failure rate can vary due to conditions of the machine, it is usually assumed to be constant over a defined period, especially over a period after what is often called a "run-in" period.

It should be noted that $F(t) = 1 - R(t)$ is the hazard function over the same period, representing the opportunity of failure of the system.

It is normal that engineering systems are designed to run-in parallel. There are many reasons. A common practice is to use a number of smaller units each of which can only deliver partial outcome but the totality of several (or all) units will be sufficient to cover the demand. A typical example is pump or power generation stations where several pumps or power generators are run-in parallel and connected to a network to deliver the required capacity.

The reliability of a parallel system of two units as shown in Fig. 8.1 is given by

$$R_S = 1 - (1 - R_1)(1 - R_2) = R_1 + R_2 - R_1 R_2 \tag{8.2}$$

where R_S is the reliability of the equivalent system; R_1 is the reliability of unit A_1; R_2 is the reliability of unit A_2.

In general, for a system with n units $A_1, A_2, \ldots, A_n$, reliability of the system is given by

$$R_S = 1 - \prod_{i=1}^{n}(1 - R_i) \tag{8.3}$$

However, Eq. (8.3) is true if a single unit is sufficient to provide the required functionality. What happens if at least two or may be more units are required to be operating at any time? This situation can be analyzed with Binominal system formulation. We use an example to illustrate the analysis process.

A water treatment plant has six 220 kW motor each driving a pump delivering 1.5 ML/h of water to residents. The average demand for treated water is 7.5 ML/h. At full load, reliability of the motor can be estimated by failure

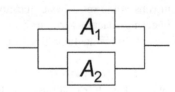

FIGURE 8.1 A parallel system with two units.

rate $= 0.1$ per 1000 h. Hence, it is possible to run five of the six pumps to fulfill the demand. If the pumps are labeled as P_1, P_2, ..., P_6, and assuming they are identical in performance, the following set of combination of pumps are possible:

$$\{(P_1, P_2, P_3, P_4, P_5), (P_1, P_2, P_3, P_4, P_6), (P_1, P_2, P_3, P_5, P_6), (P_1, P_2, P_4, P_5, P_6),$$
$$(P_1, P_3, P_4, P_5, P_6), (P_2, P_3, P_4, P_5, P_6), (P_1, P_2, P_3, P_4, P_5, P_6)\}$$

This means there are six possible combinations of five pumps running plus the possibility of one pump fails which means all six pumps are utilized. This situation is modeled by the Binominal probability theory. In general, if the probability of success of a unit is p, the combinations of an event with r units out of a total of n units (where $n > r$) are given by

$$P_r = \sum_{i=r}^{n} \frac{n!}{r!(n-r)!} p^r (1-p)^{n-r} \tag{8.4}$$

The reliability of an individual pump R over 1000 h of operation is given by

$$R = \exp(-\lambda t) \tag{8.5}$$

where λ in this example is 0.1 per 1000 h.

which gives $R = 0.9048$ for the given λ running for 1000 h. Hence, the reliability of five of six pumps system is

$$R_S = \frac{6!}{5!(6-5)!} R^5 (1-R)^{(6-5)} + \frac{6!}{6!(6-6)!} R^6 (1-R)^{(6-6)} \tag{8.6}$$

This gives an overall system reliability $R_S = 0.8957$ for the given pump system.

However, it is possible to run the pumps with a different strategy. By using variable speed drives, the pumps can be run at lower speed to save energy. Pump design theory shows that the power consumed by a pump in percent of full load power is proportional to the third power (cube) of water flow rate, i.e.,

$$x = \frac{Q^3}{Q_0^3} \tag{8.7}$$

where Q_0 is the water flow rate at full load, and Q is the water flow rate at reduced speed. It is noted that the full load run strategy has 100% constant load, i.e., 1.1 MW (220 kW $\times$ 5) power requirement.

Furthermore, if the pump (and its motor) is operated at less than full load, the failure rate (failures per 1000 h) can be reduced. Assuming that the failure rate varies according to the relationship:

$$\lambda = 0.08x + 0.02 \tag{8.8}$$

Note that when $x = 100\%$, $\lambda = 0.1$.

If the pumps are operating at a reduced load, all six pumps must be running to keep up with the demand. In case if one of the pumps fails, the remaining five pumps must then run full load to cover the demand. The reliability of the pumps then varies according to Eq. (8.6) due to the duration of full load.

Assuming that the pumps are evenly loaded in the reduced load configuration, the reliability of individual pumps is given by

$$R = \exp[-t \times (0.08x + 0.02)] \tag{8.9}$$

where t is the system's operating time. If t_r is the repair time for a pump if it fails, the system reliability depends on two components:

1. Reliability when the pumps are running reduced load, i.e., $R_{6\text{ pumps}} = \{\exp[-(1000 - t_r)(0.08x + 0.02)]\}^6$
2. Reliability when five pumps are running full load during the emergency period, i.e., $R_{5\text{ pumps}} = \{\exp[-\lambda t_r]\}^5$

The system reliability is then given by

$$R_S = R_{6\text{ pumps}} + (1 - R_{6\text{ pumps}})R_{5\text{ pumps}} \tag{8.10}$$

Assuming that $t_r = 168$ h (24 h per day for 1 full week), the combined system reliability is 0.9773.

The average power requirement for this strategy over an operating time of 1000 h (including t_r hours of one pump's repair time) is given by

$$P_S = \frac{6 \times (1000 - t_r) + 5t_r}{1000} Q_0 \tag{8.11}$$

Since six pumps are running, the load for each pump is 1.25 ML/h. The power level is then computed by Eq. (8.7) as 0.5787, and the power is 820.36 MWh. Therefore the strategy of running all pumps at reduced speed/power scenario can save energy.

It is interesting to examine the effect of change of demand to the reliability and power consumption. If the demand is at an average of 6.0 ML/h, again, two strategies can be formulated:

1. Run four pumps at full load. The other two pumps are standby. The failed pump will be repaired as soon as possible.
2. Run six pumps at reduced load. If one of the pumps fails, run five pumps at reduced load matching the demand. The failed pump will be repaired as soon as possible. The system will resume the same level for all pumps after repair.

For strategy (1), the binomial distribution theory is formulated according to four of six pumps. Using the reliability function in Eq. (8.11) of

$R = 0.9048$ for the given λ, and the system reliability Eq. (8.12), the reliability of four of six pumps system is

$$R_S = \frac{6!}{4!(6-4)!}R^4(1-R)^{(6-4)} + \frac{6!}{5!(6-5)!}R^5(1-R)^{(6-5)} + \frac{6!}{6!(6-6)!}R^6(1-R)^{(6-6)}$$

$$(8.12)$$

This gives an overall system reliability $R_S = 0.9862$ for the given pumps. Power consumed for strategy (1) is 0.88 MW/h.

For strategy (2), assuming equal load among the pumps, the reliability of the pumps can be given by

$$R_{6\ pumps} = \left\{ \exp\left[-(1000-t_r)\left(0.08\left(\frac{6.0}{6}\right)^3 + 0.02 \right) \right] \right\}^6 \qquad (8.13)$$

When one of the pumps fails, the reliability of five pumps over the emergency period is given by

$$R_{5\ pumps} = \left\{ \exp\left[-t_r\left(0.08\left(\frac{6.0}{5}\right)^3 + 0.02 \right) \right] \right\}^5 \qquad (8.14)$$

Hence, the combined system reliability can be calculated from Eq. (8.14) to be 0.9885 for $t_r = 168$ h. The power consumed over 1000 h is calculated from Eq. (8.11) to be 420.02 MWh.

The same principle can be applied to other demand situations. Table 8.1 presents the summary of the numerical analysis for some demand levels.

The two computed outcomes are plotted in Fig. 8.2 for reference. It can be seen that the reliability of the system is generally very well maintained until the capacity limit is reached when demand is closer to the maximum capacity of the pumps. In that case, there is little spare capacity for the system to manipulate and any failure will result in system breakdown.

Although Fig. 8.2 shows that there is great advantage to use strategy 2 to reduce power consumption and increase reliability, it is necessary to understand that speed reduction strategy in pump operation is restricted by the fluid dynamics in the pump. Significant deviation will violate the original design characteristics and can result in problems such as cavitation, leaking, and turbulence in the system. Since the pumps are designed with rated capacity at 7.5 ML/h, viability of the points below 6.0 ML/h in Fig. 8.2 should be verified with the actual pump characteristics.

8.3 RELIABILITY-BASED OPERATION

Design of maintenance schedule is an important engineering activity to ensure continuous delivery of system functionality to the user. The so-called reliability-based maintenance is a process that allows the maintenance

TABLE 8.1 Reliability and Power Consumption at Different Demand Levels

Demand (ML/h)	1.5	3.0	4.5	6.0	7.5	9.0
Strategy (1)						
No. of pumps	1	2	3	4	5	6
Reliability of n out of 6	1.0000	1.0000	0.9989	0.9862	0.8951	0.5488
Power consumed (MWh) (1000 h)	220	440	660	880	1100	1320
Strategy (2)						
If we run six pumps, Q for each pump	0.25	0.50	0.75	1.00	1.25	1.50
Power for Q (percent of full load)	0.0046	0.0370	0.1250	0.2963	0.5787	1.0000
Failure rate at %load	0.0204	0.0230	0.0300	0.0437	0.0663	0.1000
Repair time (h)	168	168	168	168	168	168
Reliability in 1000—repair hours (168)	0.9832	0.9811	0.9753	0.9643	0.9463	0.9202
Reliability of 6 in %load	0.9033	0.8917	0.8609	0.8040	0.7182	0.6070
One Pump Fails, Other Pumps Sharing						
Q for each pump	0.30	0.60	0.90	1.20	1.50	N/A
Power for Q (percent of full load)	0.0080	0.0640	0.2160	0.5120	1.0000	N/A
Failure rate at %load	0.0206	0.0251	0.0373	0.0610	0.1000	N/A
Reliability of pump in emergency load	0.9965	0.9958	0.9938	0.9898	0.9833	0.9833
Reliability of 5 in full load for 1 week	0.9828	0.9791	0.9692	0.9501	0.9194	0.0000
Combined reliability	0.9983	0.9977	0.9957	0.9902	0.9773	0.6070
Power consumed (MWh) (1000 h operation)	6.56	52.50	177.20	420.02	820.36	1320.00

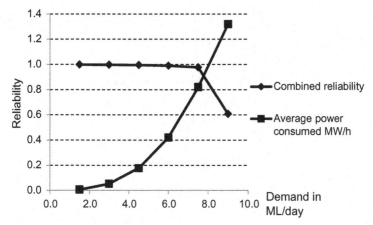

FIGURE 8.2   Reliability and power consumption at different demand levels.

engineer to adjust, or more precisely, optimize the maintenance schedule to reduce cost of ownership. Similar philosophy can be applied to system operation aiming to eliminate downtime as much as possible. The basis of this type of computation relies on the reliability function, often expressed as Weibull distribution.

The Weibull distribution takes the following form:

$$f(t) = \frac{\beta t^{\beta-1}}{\eta^{\beta}} \exp\left[-\left(\frac{t}{\eta}\right)^{\beta}\right] = \frac{\beta}{\eta}\left(\frac{t}{\eta}\right)^{\beta-1} \exp\left[-\left(\frac{t}{\eta}\right)^{\beta}\right] \tag{8.15}$$

where $t \geq 0$, $\beta > 0$, $\eta > 0$

The reliability function is given by

$$R(t) = \int_{t}^{\infty} f(t)dt = \exp\left[-\left(\frac{t}{\eta}\right)^{\beta}\right] \tag{8.16}$$

and the cumulative failure function:

$$Q(t) = 1 - R(t) = 1 - \exp\left[-\left(\frac{t}{\eta}\right)^{\beta}\right] \tag{8.17}$$

The factor β is known as shape parameter. Fig. 8.3 shows the shape of the failure rate function changes with different values of β.

The factor η is known as scale parameter. It can be seen from Fig. 8.3 that if $\beta = 1$, the function is constant at $(1/\eta)$. η is the time at which 63.2% of the components fail.

An important property of the Weibull distribution is that as β increases, the mean of the distribution approaches η and the variance approaches zero.

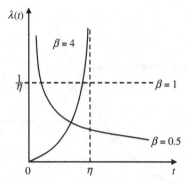

FIGURE 8.3 Effect of shape parameter.

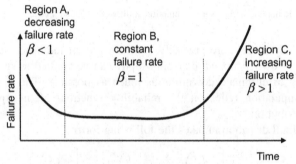

- Region A – running in period
- Region B – normal operation
- Region C – deterioration, due to wear out

FIGURE 8.4 The Bathtub curve.

η and β change the shape of the distribution making the Weibull distribution a good tool for matching a wide range of failure data patterns found in engineering practice as shown in Fig. 8.4. The complex curve in Fig. 8.4 is often known as the Bathtub curve.

The Weibull factors β and η are often found from fitting historical data to the Weibull distribution. There are different means of doing this. The Bernard's Median Rank method estimates the reliability function by the number of failure records in a population with total number of units N:

$$R(t) = 1 - \frac{i - 0.3}{N + 0.4} \tag{8.18}$$

where i is the ith failure

For a system that requires k machines to work simultaneously, the reliability of the parallel system $R(t)$ is given by

$$R(t) = R_1(t) \cdot R_2(t) \cdots R_k(t) \tag{8.19}$$

TABLE 8.2 Weibull Factors for Gas Turbines Daily Reliability

Entity	β	η
Gas turbine G1	1.6	85
Gas turbine G2	1.7	80
Gas turbine G3	1.4	75
Gas turbine G4	1.5	80

Using Eq. (8.16) the Weibull equation can be linearized by taking natural logarithm twice:

$$\ln\{ - \ln[R(t)]\} = \beta \cdot \ln(t) + [- \beta \cdot \ln(\eta)] \qquad (8.20)$$

where t is the time of failure. The β term at the right hand side is the slope of the regression line, while the term $-\beta \cdot \ln(\eta)$ is the Y-intercept. We use an example to illustrate how the Weibull distribution can be used to determine an optimal operation strategy.

A large complex commercial center is keen to provide most reliable power supply system to its tenants. The center has four gas turbines each of capacity 10 MW but their reliability is different. The gas turbines are connected to the main grid by switches in standby mode. Table 8.2 presents the Weibull factors of the gas turbines after the most recent review of their historical daily performances.

The demand for power in MW varies as in a 24-h daily cycle with Eq. (8.21):

$$P = P_a - P_o \sin 2\pi \left(\frac{8 - t}{24} \right) \qquad (8.21)$$

where P_a is the Base power and P_o is the average power.

The demand must be met by the gas turbines, i.e.,

$$10n \geq P \qquad (8.22)$$

where n is the number of active gas turbines.

The gas turbines do not need to run all the time. Some periods in the day can be supplied by one, two, or three gas turbines only. To meet the demand the system is set up such that it will review the power requirement every review period. This will incur some work to set up the system and is a fixed cost for each period. Let us denote this cost as \$$p$ per review.

The other cost that should be considered in the analysis is the cost of the breakdown. This cost includes loss of revenue, compensation claims, possible overtime pay, emergency maintenance work, etc. The cost is proportional to hours of breakdown. Let us denote this cost as \$$b$ per hour.

Hence, the total daily running cost is given by

$$C_m = (1 - R_m)bt + \frac{24}{t}p \quad \text{where } t = \text{review period duration} \qquad (8.23)$$

At any time, if a gas turbine is not required, it will be stopped and serviced. Restarting of a gas turbine to the state capable of switching into the supply grid will take 20 min. This means if there is a failure of one of the running gas turbine, it is inevitable a black out of at least 20 min will occur, and this should be minimized as much as possible. Hence, the principle for minimizing the expected black out is to run the combination of gas turbines with the highest reliability, i.e., for each combination of n gas turbines

$$R_m = \max(R_i) \quad \text{where } i = \text{the set of binomial combination of} \atop n \text{ out of four gas turbines} \qquad (8.24)$$

To evaluate the strategies, some numerical values are given. The cost of review is \$10 per review. A perfect switching system for connecting gas turbines power to the grid is assumed.

If the cost of breakdown is \$1000 per hour, what is the best strategy to operate these gas turbines in any day of operation?

The reliability of the combination of gas turbines in an hour can be computed from Eq. (8.23) for single gas turbine and from Eq. (8.24) for multiple gas turbines, as presented in Table 8.3. The maximum reliability combination for each combination is bolded.

Since the demand varies in the day, the operating schedule will need to be planned according to much shorter period within the day. If we plan by the hour, the number of gas turbines that must be operating can be computed in Table 8.4. Hence, the reliability of the system depends on the highest assessed reliability of the combination of gas turbines.

The variation of reliability can be shown graphically in Fig. 8.5.

The average reliability over the 24 h period is given by $E(R(t)) = 0.9759$.

To investigate the effect of review period to the running cost and systems reliability, the same model is used to generate Table 8.5 for $t = 0.5$ h. It should be noted that due to change of review period, the reliability of gas turbines for the next half hour operation is presented in Table 8.6.

The variation of reliability can be shown graphically in Fig. 8.6.

The basic pattern of variations in system reliability is similar but the reliability values have improved. The average reliability over the 24 h period is given by $E(R(t)) = 0.9917$.

The modeling process repeats for other review durations. It is necessary to point out that for review periods shorter than 20 min, the break down time is set at the minimum of 20 min to restart a gas turbine. The system reliability and daily cost are tabulated in Table 8.7 and graphically in Figs. 8.7 and 8.8.

TABLE 8.3 Reliability of Gas Turbine Operations for the Next Hour

Gas Turbine Operations	$R(t)$, $t = 1$
Gas turbine 1 only	0.9920
Gas turbine 2 only	**0.9934**
Gas turbine 3 only	0.9827
Gas turbine 4 only	0.9882
GT1 and GT2	**0.9855**
GT1 and GT3	0.9748
GT1 and GT4	0.9803
GT2 and GT3	0.9762
GT2 and GT4	0.9817
GT3 and GT4	0.9711
GT1, GT2, GT3	0.9684
GT1, GT2, GT4	**0.9739**
GT2, GT3, GT4	0.9647
GT1, GT3, GT4	0.9633
All gas turbines	**0.9570**

It can be seen from Table 8.7 that the minimum expected daily cost is estimated to occur at a review period duration of 0.6. The average reliability curve is always decreasing.

8.4   MARKET VARIATION MODELING

Commercial market changes frequently and irregularly. It is difficult, if not impossible, to predict how a company or a product can perform in certain market. On the other hand, to make a decision on an investment, irrespective of whether it is an enhancement of existing products or development of a completely new product, some kind of market variation analysis should be conducted to guide the process.

This type of systems analysis involves modeling of present and future costs, and hence develops a view of the financial risks or profitability of the investment. The steps involved in investment appraisal are

1. Collect information about the investment project
2. Develop system model
3. Calculate return
4. Make decision

TABLE 8.4 The Combination of Gas Turbines With the Highest Reliability at Different Hours of the Day

Hour of day	0	1	2	3	4	5	6	7	8	9	10	11	12
Power demand	6.412	4.613	4.000	4.613	6.412	9.272	13.000	17.341	22.000	26.659	31.000	34.728	37.588
No. of gas turbines in operation	1	1	1	1	1	1	2	2	3	3	4	4	4
Reliability of the system	0.9934	0.9934	0.9934	0.9934	0.9934	0.9934	0.9855	0.9855	0.9739	0.9739	0.9570	0.9570	0.9570
Cost per break down ($)	6.5608	6.5608	6.5608	6.5608	6.5608	6.5608	14.504	14.504	26.149	26.149	43.027	43.027	43.027

Hour of day	13	14	15	16	17	18	19	20	21	22	23	24	Expected daily cost (inc. review)
Power demand	39.387	40.000	39.387	37.588	34.728	31.000	26.659	22.000	17.341	13.000	9.2721	6.4115	
No. of gas turbines in operation	4	4	4	4	4	4	3	3	2	2	1	1	
Reliability of the system	0.9570	0.9570	0.9570	0.9570	0.9570	0.9570	0.9739	0.9739	0.9855	0.9855	0.9934	0.9934	
Cost per break down ($)	43.027	43.027	43.027	43.027	43.027	43.027	26.149	26.149	26.149	26.149	6.5608	6.5608	$842.34

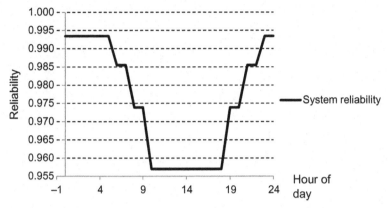

FIGURE 8.5   Reliability of gas turbines combination planned every hour.

TABLE 8.5 Reliability of Gas Turbine Operations for the Next Half Hour

Gas Turbine Operations	$R(t)$, $t = 0.5$
Gas turbine 1 only	0.9974
Gas turbine 2 only	**0.9980**
Gas turbine 3 only	0.9934
Gas turbine 4 only	0.9958
GT1 and GT2	**0.9953**
GT1 and GT3	0.9908
GT1 and GT4	0.9932
GT2 and GT3	0.9914
GT2 and GT4	0.9938
GT3 and GT4	0.9892
GT1, GT2, GT3	0.9888
GT1, GT2, GT4	**0.9912**
GT2, GT3, GT4	0.9872
GT1, GT3, GT4	0.9866
All gas turbines	**0.9846**

In step 1, any data related to the investment should be collected. It is not possible to categorically specify what information is required for which project. However, common sense should always apply, i.e., any data can be useful and should be captured with as much detail as possible, e.g., time,

TABLE 8.6 The Combination of Gas Turbines With the Highest Reliability at Different Half Hours of the Day

	0.0	0.5	1.0	1.5	2.0	2.5	3.0	3.5	4.0	4.5	5.0	5.5	6.0
Half hour of day	0.0	0.5	1.0	1.5	2.0	2.5	3.0	3.5	4.0	4.5	5.0	5.5	6.0
Power demand	6.412	5.370	4.613	4.154	4.000	4.154	4.613	5.370	6.412	7.720	9.272	11.042	13.000
No. of gas tur.	1	1	1	1	1	1	1	1	1	1	1	2	2
Reliability	0.9980	0.9980	0.9980	0.9980	0.9980	0.9980	0.9980	0.9980	0.9980	0.9980	0.9980	0.9953	0.9953
Cost per b/d ($)	1.012	1.012	1.012	1.012	1.012	1.012	1.012	1.012	1.012	1.012	1.012	2.3316	2.3316
Half hour of day	6.5	7.0	7.5	8.0	8.5	9.0	9.5	10.0	10.5	11.0	11.5	12.0	
Power demand	15.112	17.341	19.651	22.000	24.349	26.659	28.888	31.000	32.958	34.728	36.280	37.588	
No. of gas tur.	2	2	2	3	3	3	3	4	4	4	4	4	
Reliability	0.9953	0.9953	0.9953	0.9912	0.9912	0.9912	0.9912	0.9846	0.9846	0.9846	0.9846	0.9846	
Cost per b/d ($)	2.3316	2.3316	2.3316	4.4187	4.4187	4.4187	4.4187	7.691	7.691	7.691	7.691	7.691	
Half hour of day	12.5	13.0	13.5	14.0	14.5	15.0	15.5	16.0	16.5	17.0	17.5	18.0	
Power demand	38.630	39.387	39.846	40.000	39.846	39.387	38.630	37.588	36.280	34.728	32.958	31.000	
No. of gas tur.	4	4	4	4	4	4	4	4	4	4	4	4	

Reliability	0.9846	0.9846	0.9846	0.9846	0.9846	0.9846	0.9846	0.9846	0.9846	0.9846	0.9846	0.9846	
Cost per b/d ($)	7.691	7.691	7.691	7.691	7.691	7.691	7.691	7.691	7.691	7.691	7.691	7.691	
Half hour of day	18.5	19.0	19.5	20.0	20.5	21.0	21.5	22.0	22.5	23	23.5	24	Expected daily cost
Power demand	28.888	26.659	24.349	22.000	19.651	17.341	15.112	13.000	11.042	9.272	7.720	6.412	
No. of gas tur.	3	3	3	3	2	2	2	2	2	1	1	1	
Reliability	0.9912	0.9912	0.9912	0.9912	0.9953	0.9953	0.9953	0.9953	0.9953	0.9980	0.9980	0.9980	
Cost per b/d ($)	4.4187	4.4187	4.4187	4.4187	2.3316	2.3316	2.3316	2.3316	2.3316	1.012	1.012	1.012	$683.58

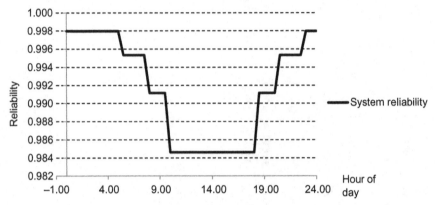

FIGURE 8.6   Reliability of gas turbines combination in half hour planning cycle.

TABLE 8.7 Summary of Daily Cost and Reliability of Gas Turbine Operations According to Review Interval

Review Interval	Expected Daily Cost ($)	Average Reliability
1.000	842.336	0.9759
0.750	701.720	0.9846
0.650	677.914	0.9875
0.600	672.206	0.9889
0.550	682.149	0.9902
0.500	683.579	0.9917
0.450	715.967	0.9929
0.400	745.475	0.9940
0.333	829.867	0.9955
0.300	903.507	0.9962
0.275	979.708	0.9966
0.250	1053.507	0.9971
0.225	1159.330	0.9975
0.200	1284.168	0.9979
0.150	1671.730	0.9987
0.100	2458.164	0.9993

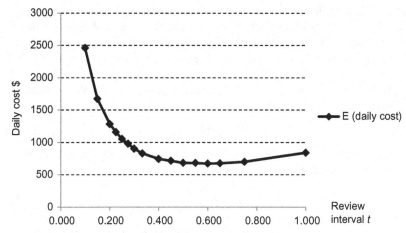

FIGURE 8.7 Numerical determination of expected daily cost function.

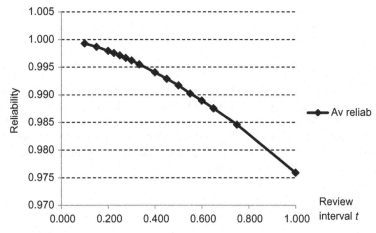

FIGURE 8.8 Numerical determination of reliability function.

source, circumstance, accuracy, etc. In many cases the data can come from third party sources such as bureau of statistics, company reports, financial research.

Step 2 is probability the most critical part of the investment appraisal process. Investment modeling usually employs the concept of time value of money. This means the data from step 1 will need to be converted to a monetary term, if the data are not initially collected as such. In numerical analysis, so long as the modeling is logically put together to capture variations in valuation of data, a sensitivity analysis can always be performed later to ascertain a level of confidence in the outcomes.

In evaluating alternative investments, we are concerned with the economic impact of cash flows over time. Extra money will be accumulated by value added activities over time. In such problems, the time value of money, that is, the impact of time on cash flows must be recognized. The decision to invest is based on the prospect of future cash flows of sufficient size to repay the original investment and give the investor a reasonable profit. In Fig. 8.9, the principle P is invested into the project, which will go through n periods to produce a future return F. If the financial interest is $i\%$, the relationship between P and F is given by

$$F = P(1+i)^n \tag{8.25}$$

The term $(1 + i)^n$ is called the future worth factor of a present amount. Eq. (8.25) is often rearranged as

$$P = F\frac{1}{(1+i)^n} \tag{8.26}$$

The term $1/(1 + i)^n$ is called the present worth factor of a future amount.

The financial interest i is often known as the marginal attractive rate of return (MARR). It is a rate of return that the investor is attracted to invest into this project. If the rate of return is lower than i, the investor will not be attracted and hence it is a demarcation line of decision—at the decision margin.

In any investment project, it is normal that there are additional funding injected into the project at a time other than the initiation stage, e.g., sales revenue. Likewise, it is normal that dividends may be taken out of the project at different times. This situation can be represented in Fig. 8.10. Without loss of generality, the injection and withdrawal of funds (or values) are denoted to occur at the end of each period, and only one injection and one withdrawal are denoted in that period.

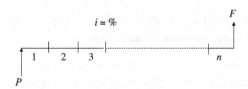

FIGURE 8.9 Time value of principal and future return.

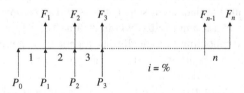

FIGURE 8.10 Investment values can change over the analysis period.

There are several ways of analyzing this situation. The net present value (NPV) method converts all proceeds of investment changes to the present time using Eq. (8.26). The situation in Fig. 8.9 can then be represented by

$$\text{NPV} = \sum_{j=1}^{n} \frac{F_j - P_j}{(1+i)^j} - P_0 \tag{8.27}$$

It is noted that the future worth method is basically the same as NPV method except that Eq. (9.25) is used instead.

In real life the data can never be exactly estimated. There is always uncertainty in market survey data, for example, sales in good, fair, and poor market conditions. If the funding changes can be represented by a variable $X = \{x_1, x_2, \ldots, x_i, \ldots, x_m\}$ with probabilities for each sample space $p(x_k) = p_k$ is given by

$$E(X) = \sum_{k=1}^{m} x_k p_k \text{ where } \sum_{k=1}^{m} p_k = 1 \tag{8.28}$$

Eq. (8.28) can then be modified as

$$\text{NPV} = \sum_{j=1}^{n} \frac{\sum_{k=1}^{m}(F_{jk} - P_{jk})p_k}{(1+i)^j} - P_0 \tag{8.29}$$

If a project is worth investing, the NPV in Eq. (8.29) should be zero or positive with a given value of i.

Once the system model is developed using the above concepts and equations, step 3 is to calculate the NPV for the project. To illustrate how the system model works, we use an example.

A company is considering building a new plant for a new product although the future success of this product is uncertain. The development cost of the product is $10 M. The sales volume (in $M) depends on the market conditions W1, W2, and W3. The probability of W1, W2, and W3 to occur is presented in Table 8.8. The investment duration is set to be 10 years.

The sale volumes (in M units) for each of the market conditions (due to competition, alternatives, economic) are presented in Table 8.9.

TABLE 8.8 Probability of Different Market Conditions

Market condition	W1	W2	W3
Probability	0.4	0.3	0.3

TABLE 8.9 Sale Volume (in M units) for Different Market Conditions

Year	1	2	3	4	5	6	7	8	9	10
W1	1.3	2.1	2.9	3.8	2.2	1.4	2.5	3.6	4.2	3.7
W2	0.1	1.0	1.8	1.2	2.1	1.1	0.9	0.8	1.1	0.3
W3	2.2	4.3	3.4	3.8	2.9	5.2	2.9	4.4	3.3	4.2

Two options of plant design can be built. A large plant can produce 3 M (million) units per year and will cost \$20 M. A small plant can produce 1 M (million) units per year and will cost \$8 M. Both options can be operated immediately, i.e., in time for sales in year 1. The large and small plants will have a residual value of \$2 M and \$0.5 M at the end of the project duration.

Other information collected:

- Material cost \$1 per unit.
- Labor cost \$0.5 per unit.
- Company overhead 200%.
- Sale price \$8 per unit.
- MARR 10%.
- The product is expected to have a shelf life of only 3 years. If the shelf life expires, the product has to be scrapped.
- Lost sale in the year cannot be compensated in coming years.

Two questions to be answered:

1. Is this project worth investing?
2. Which option of plant should be built?

The decision for each option of plant should be built depends on the NPV for the option. To calculate the NPV, Table 8.10 is setup for the case of large plant and market condition W1.

The rows are computed as follows:

- Product development and large plant are entered as outgoing funds (negative).
- Product volume is fixed by the use of large plant.
- The yearly stock rows are computed starting from 3-year stock: for year k,

$$y_{3k} = if \begin{cases} y_{2(k-1)} < 0, & y_{3k} = 0.0 \\ y_{2(k-1)} \geq 0, & y_{3k} = q_k - y_{2(k-1)} \end{cases}$$

- The 2-year stock rows are then computed:

$$y_{2k} = if \begin{cases} y_{1k} > 0, & y_{2k} = y_{1(k-1)} \\ y_{1k} \leq 0, & y_{2k} = y_{1(k-1)} + y_{3k} \end{cases}$$

TABLE 8.10 Computation of the Case Large Plant With Market Condition W1

Year	0	1	2	3	4	5	6	7	8	9	10
Product development ($M)	−10										
Large plant ($M)	−20										
Production volume (M units) q_i		3.0	3.0	3.0	3.0	3.0	3.0	3.0	3.0	3.0	3.0
Sales under W1 s_i		1.3	2.1	2.9	3.8	2.2	1.4	2.5	3.6	4.2	3.7
One-year stock (M units) y_{1i}	0.0	1.7	2.6	3.0	3.0	3.0	3.0	3.0	3.0	3.0	3.0
Two-year stock (M units) y_{2i}			−0.4	2.6	1.8	2.6	3.0	3.0	2.4	1.2	0.5
Three-year stock (M units) y_{3i}				0.0	−1.2	−0.4	1.2	0.5	−0.6	−1.8	−2.5
Actual sale (units) a_i		1.3	2.1	2.9	3.8	2.2	1.4	2.5	3.6	4.2	3.7
Sale revenue ($M)		10.4	16.8	23.2	30.4	17.6	11.2	20.0	28.8	33.6	29.6
Material cost ($M)		3.0	3.0	3.0	3.0	3.0	3.0	3.0	3.0	3.0	3.0
Labor cost ($M)		1.5	1.5	1.5	1.5	1.5	1.5	1.5	1.5	1.5	1.5
Overhead cost ($M)		3.0	3.0	3.0	3.0	3.0	3.0	3.0	3.0	3.0	3.0
Profit ($M)		2.9	9.3	15.7	22.9	10.1	3.7	12.5	21.3	26.1	22.1
PV ($M)		2.636	7.686	11.796	15.641	6.271	2.089	6.414	9.937	11.069	8.521
NPV ($M)	52.059										

- The 1-year stock rows are then computed:

$$y_{1k} = if \begin{cases} y_{2k} < 0, & y_{1k} = q_k + y_{2k} \\ y_{1k} \geq 0, & if \begin{cases} y_{1(k-1)} < 0, & y_{1k} = q_k + y_{2k} - s_k \\ y_{1(k-1)} \geq 0, & q_k \end{cases} \end{cases}$$

- The actual sales volume is computed from: $a_k = if \begin{cases} y_{1k} > 0, & a_k = s_k \\ y_{1k} \leq 0, & a_k = q_k \end{cases}$

- Sales revenue, material and labor cost are computed from $8, $1, and $0.5 per unit respectively multiplied with actual sales volume.
- Overhead cost is computed from 200% of labor cost.
- Profit = sales revenue − material cost − overhead cost.
- PV is computed from Eq. (8.26).
- NPV is the sum of all PVs in the table.

Similarly, Table 8.11 is repeated for market conditions W2 and W3 in Table 8.12.

The same process is repeated for small plant. The key figure from these tables is the NPV. Table 8.13 presents the summary for all six tables.

Step 4 of the project appraisal process is to recommend a decision. It is clear from Table 8.13 that large plant has a higher expected NPV for this investment. Furthermore the expected NPV is positive which means the project is worth investing. Without other influential factors, building a large plant is recommended. However, decision making is not a straightforward process. The management board should also consider other factors such as availability of funds, validity of the market survey, operation support, and human resources.

Once the model is setup, it is then possible to examine the effect of changing i on the NPVs. Fig. 8.11 plots the expected NPV of large plant against different MARR values. It can be seen that the expected NPV is zero somewhere between 29% and 30%.

To find a more accurate x-intercept value the bisection method can be used. From the data table of Fig. 8.10 a zero occurs between internal rate of return (IRR) values of 29% and 30%. Applying bisection method continuously until the NPV of large plant is less than a tolerable value, say, less than $0.01 M. The computation results are presented in Table 8.14.

Table 8.14 is plotted in Fig. 8.12 to show the convergence process.

The x-intercept value is found to be 29.3457%. This value is known as IRR. This is the inherent percentage of return that this project can incorporate of what the investor expects, i.e., when the NPV value of the project is zero. Since this value is higher than the investor's anticipated MARR of 10%, the project is worth investing.

8.5 FAILURE MODE AND EFFECT ANALYSIS PRIORITIZATION

In the manufacturing world the reliability of products is very critical to the outcome of the functionality of the finished product. In order to ensure that

TABLE 8.11 Computation of the Case Large Plant With Market Condition W2

Year	0	1	2	3	4	5	6	7	8	9	10
Product development ($M)	−10										
Large plant ($M)	−20										
Production volume (M units) q_i		3.0	3.0	3.0	3.0	3.0	3.0	3.0	3.0	3.0	3.0
Sales under W1 (M units) s_i		0.1	1.0	1.8	1.2	2.1	1.1	0.9	0.8	1.1	0.3
One-year stock (M units) y_{1i}	0.0	2.9	3.0	3.0	3.0	3.0	3.0	3.0	3.0	3.0	3.0
Two-year stock (M units) y_{2i}			1.9	3.0	3.0	3.0	3.0	3.0	3.0	3.0	3.0
Three-year stock (M units) y_{3i}				0.1	1.8	0.9	1.9	2.1	2.2	1.9	2.7
Actual sale (units) a_i		0.1	1.0	1.8	1.2	2.1	1.1	0.9	0.8	1.1	0.3
Sale revenue ($M)		0.8	8.0	14.4	9.6	16.8	8.8	7.2	6.4	8.8	2.4
Material cost ($M)		3.0	3.0	3.0	3.0	3.0	3.0	3.0	3.0	3.0	3.0
Labor cost ($M)		1.5	1.5	1.5	1.5	1.5	1.5	1.5	1.5	1.5	1.5
Overhead cost ($M)		3.0	3.0	3.0	3.0	3.0	3.0	3.0	3.0	3.0	3.0
Profit ($M)		−6.7	0.5	6.9	2.1	9.3	1.3	−0.3	−1.1	1.3	−5.1
PV ($M)		−6.091	0.413	5.184	1.434	5.775	0.734	−0.154	−0.513	0.551	−1.966
NPV ($M)	−24.633										

TABLE 8.12 Computation of the Case Large Plant With Market Condition W3

Year	0	1	2	3	4	5	6	7	8	9	10
Product development ($M)	−10										
Large plant ($M)	−20										
Production volume (M units) q_i		3.0	3.0	3.0	3.0	3.0	3.0	3.0	3.0	3.0	3.0
Sales under W1 (M units) s_i		2.2	4.3	3.4	3.8	2.9	5.2	2.9	4.4	3.3	4.2
One-year stock (M units) y_{1i}	0.0	0.8	−0.5	−0.4	−0.8	0.1	−2.1	0.1	−1.3	−0.3	−1.2
Two-year stock (M units) y_{2i}			−3.5	0.0	0.0	0.0	−5.1	0.0	−4.3	0.0	0.0
Three-year stock (M units) y_{3i}				0.0	0.0	0.0	0.0	0.0	0.0	0.0	0.0
Actual sale (units) a_i		2.2	3.8	3.0	3.0	2.9	3.1	2.9	3.1	3.0	3.0
Sale revenue ($M)		17.6	30.4	24.0	24.0	23.2	24.8	23.2	24.8	24.0	24.0
Material cost ($M)		3.0	3.0	3.0	3.0	3.0	3.0	3.0	3.0	3.0	3.0
Labor cost ($M)		1.5	1.5	1.5	1.5	1.5	1.5	1.5	1.5	1.5	1.5
Overhead cost ($M)		3.0	3.0	3.0	3.0	3.0	3.0	3.0	3.0	3.0	3.0
Profit ($M)		10.1	22.9	16.5	16.5	15.7	17.3	15.7	17.3	16.5	16.5
PV ($M)		9.182	18.926	12.397	11.270	9.748	9.765	8.057	8.071	6.998	6.361
NPV ($M)	70.774										

TABLE 8.13 Summary of Plant and Market Conditions

Plant	Market Condition	NPV	Probability	Expected NPV ($M)
Large	W1	52.059	0.4	34.66605
	W2	− 24.633	0.3	
	W3	70.774	0.3	
Small	W1	15.795	0.4	14.54492
	W2	11.628	0.3	
	W3	15.795	0.3	

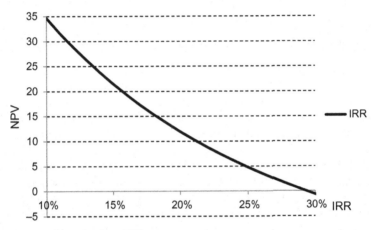

FIGURE 8.11   Determination of IRR.

TABLE 8.14 Bisection Steps to Find Zero of NPV of Large Plant

Iteration	MARR Used	NPV of Large Plant ($M)
1	29.500%	− 0.150409
2	29.250%	0.094999
3	29.375%	− 0.028094
4	29.313%	0.033350
5	29.344%	0.002610
6	29.359%	− 0.012750
7	29.352%	− 0.005074
8	29.348%	− 0.001234
9	29.346%	0.000686

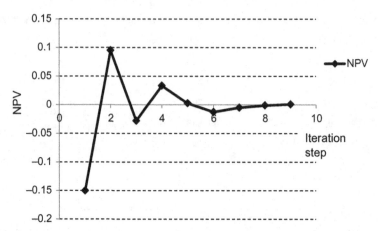

FIGURE 8.12   Change of NPV of large plant using bisection method.

the final product is of good quality, it is important that the production engineers involve right from the initial stages the product meets the set quality standards. Failure mode and effects analysis (FMEA) is a tool which has been found helpful in the analysis of failure mode and also the reliability.

FMEA is usually used for analyzing potential failures of products during design phase. The analysis process includes an initial stage of understanding the product's design (in terms of its components and linkages) and its operation so as to recognize possible malfunctions. Each of the malfunctions can then be explored in details to identify the effect of failures on the function of the product.

When using FMEA tool in manufacturing, the "product" is the production system. Since the system does not normally form a hard touchable system, identification of the malfunctions could be difficult. The method to determine the priorities can be described with the following steps:

1. For each of the error, a cause—effect diagram (also known as fish bone chart) is developed, normally with the involvement of personnel from quality, production and engineering departments. The cause—effect diagram will help to identify the causes of the error and will form the basis for deciding remedial actions.
2. Using the list of causes in the cause—effect diagram, perform a FMEA to determine the ranking priority numbers (RPNs), which will establish a priority list of causes to be tackled.
3. From the FMEA RPN list, select a reasonable number of top causes and create their vector of priority (VP) using analytic hierarchy process (AHP) methodology. The number of causes to be used depends on the amount of effort or resources that this exercise is going to be allocated.

4. A fault tree linking the selected causes to the operational error is then created. The causes are OR-ed to form the secondary and higher level faults because they can generate the errors independently.

The quality risk ranking method is shown diagrammatically in Fig. 8.13. It is clear that the procedure is not representable by a normal mathematical formulation. The problem needs to be analyzed by numerical means. From the procedure described earlier, several analysis tools are required for this method. The cause—effect diagram is often known as the fish bone chart because it looks like a fish bone with the head usually drawn at the right hand side (Fig. 8.14). The fish head is the problem that we would like to solve. The bones are possible causes of the problem. Each branch can be expanded further into more details for easier analysis.

The cause—effect diagram is a logic assistant tool to identify possible causes of the problem in preparation for the next step of quality risk ranking.

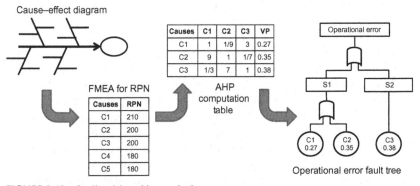

FIGURE 8.13   Quality risk ranking method.

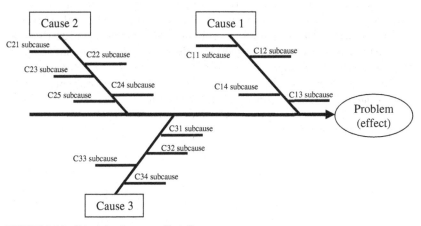

FIGURE 8.14   Principle of cause—effect diagram.

FMEA was developed in the 1950s and was one of the first systematic methods used to analyze failure in technical systems. FMEA is a rating or ranking system for examining a set of components or activities in a system and determining whether the risk associated with them is satisfactory. FMEA takes the causes and subcauses identified in the cause−effect diagram and develops into a numerical ranking hierarchy.

To complete step 2 earlier, FMEA relies on the assessment of an indicator called risk priority number (RPN), which consists of three ratings:

$$RPN = S \times O \times D \qquad (8.30)$$

where S is the severity of each effect of failure; O is the likelihood of occurrence for each cause of failure; D is the likelihood of prior detection for each cause of failure (i.e., the likelihood of detecting the problem before it reaches the end user or customer. Note: a higher value means easier to detect).

All ratings are assessed between 1 and 10, where 1 means no effect and 10 means maximum. A higher RPN indicates that the failure is critical and can be detected easily. The effect of rectifying the problem is relatively effective and hence attracts higher priority.

To prioritize remedial actions on potential failures, FMEA requires an estimate of the relative probability of occurrence of the failures. Assessment of RPN in FMEA is often subjective to the experience and attitude of the engineer assessing the system. To minimize any bias in opinion, AHP is ideal for a team of assessors to work together to develop the priority ranking.

The AHP method was developed by the American mathematician Saaty in the 1970s. AHP is used in manufacturing decision making, for example, as a strategic decision-making tool to justify machine tool selection. AHP allows for interactions and interdependence among factors in complex problems and still enables the assessors to think about these interactions in a simple way.

The AHP preference scale is based on the level of preference a given respondent has between two criteria, i.e., a pairwise comparisons. The numerical level of preferences from 1 to 9 and the definitions of that level of preference are given in Table 8.15. The numerical judgment made can also be considered as answering the question: *How many times more dominant is one element than the other with respect to a certain criterion or attribute?*

The AHP technique relies on the supposition that humans are more capable of making relative judgments than absolute judgments. Judgment matrices are created based on responses from respondents. The judgment matrix **A** can be formulated as follows:

$$\mathbf{A} = \left[a_{ij} \right] = \begin{bmatrix} \dfrac{w_1}{w_1} & \dfrac{w_1}{w_2} & \cdots & \dfrac{w_1}{w_n} \\[2mm] \dfrac{w_2}{w_1} & \dfrac{w_2}{w_2} & \cdots & \dfrac{w_2}{w_n} \\[2mm] \dfrac{\dot{w_n}}{w_1} & \dfrac{\dot{w_n}}{w_2} & \cdots & \dfrac{\dot{w_n}}{w_n} \end{bmatrix} \quad i,j = 1, \ldots, n \qquad (8.31)$$

TABLE 8.15 The AHP Preference Scale

Level of Preference	Definition	Explanation
1	Equal importance/ preference	Two activities contribute equally to the objective
3	Moderate importance/ preference	Experience and judgment slightly favor one activity over another
5	Strong importance/ preference	Experience and judgment strongly favor one activity over another
7	Very strong or demonstrated importance/preference	An activity is favored very strongly over another; its dominance demonstrated in practice
9	Extreme importance/ preference	The evidence favoring one activity over another is of the highest possible order of affirmation
2, 4, 6, 8	For compromises between the above values	Sometimes one needs to interpolate a compromise judgment numerically because there is no good word to describe it

Ideally, two numbers w_i and w_j would be known (tangibles). In general, though w_i and w_j are not known.

The values of w_i/w_j are estimated by a person or group making judgments on the extent of dominance of w_i over w_j. It is easy to see that

$$a_{ij}a_{ji} = 1 \quad i \neq j \tag{8.32}$$

$$a_{ii} = 1 \quad i = 1, \ldots, n \tag{8.33}$$

It is important to eliminate major inconsistencies in the individual's or group's reasoning. For a perfectly consistent matrix then the a_{ij} can be exactly determined from the w_i and w_j values. In the real world, consistency of judgment is seldom ideal. A measure of consistency is the Consistency Index (CI), for a matrix size n:

$$CI = \frac{\lambda_{\max} - n}{n - 1} \tag{8.34}$$

where

$$\lambda_{\max} = \sum_{j=1}^{n} a_{ij} \frac{w_j}{w_i} \tag{8.35}$$

If the matrix is perfectly consistent,

$$a_{ij} = \frac{w_i}{w_j} \quad \text{for } i, j = 1, \ldots, n \tag{8.36}$$

The CI value increases with increasing inconsistency. Studies performed using random entries for the judgment values a_{ij} have determined values for CI obtained from random inputs. These values for different matrix sizes are called the Random Consistency Index (RI). The measure of interest for consistency is the consistency ratio (CR):

$$CR = \frac{CI}{RI} \tag{8.37}$$

The recommended value is $CR \leq 0.1$ to ensure acceptable consistency. Acceptable consistency is needed as if new information is added to an inconsistent matrix the weightings may not change in a consistent fashion, and as a result the additional information may be of little value as a basis for better decisions.

The judgment matrix **A** is normalized to produce VP by the following numerical process:

1. Add each column to a column total and then divide the values in the column by the total:

$$t_j = \sum_{i=1}^{n} a_{ij} \tag{8.38}$$

$$\mathbf{B} = \begin{bmatrix} b_{ij} \end{bmatrix} = \begin{bmatrix} \dfrac{a_{11}}{t_1} & \dfrac{a_{12}}{t_2} & \cdots & \dfrac{a_{1n}}{t_n} \\[2ex] \dfrac{a_{21}}{t_1} & \dfrac{a_{22}}{t_2} & \cdots & \dfrac{a_{2n}}{t_n} \\[2ex] \dfrac{a_{n1}}{t_1} & \dfrac{a_{n2}}{t_2} & \cdots & \dfrac{a_{nn}}{t_n} \end{bmatrix} = \begin{bmatrix} \dfrac{a_{ij}}{t_j} \end{bmatrix} \tag{8.39}$$

2. Add each row to form a $n \times 1$ matrix.

$$\mathbf{C} = [c_j] = \begin{bmatrix} \sum_{j=1}^{n} b_{1j} \\[2ex] \sum_{j=1}^{n} b_{2j} \\ \cdots \\ \sum_{j=1}^{n} b_{nj} \end{bmatrix} \tag{8.40}$$

3. Normalize matrix **C** in a similar way:

$$\mathbf{D} = [d_j] = \begin{bmatrix} \dfrac{\sum\limits_{j=1}^{n} b_{1j}}{\sum\limits_{i=1}^{n}\sum\limits_{j=1}^{n} b_{ij}} \\[2ex] \dfrac{\sum\limits_{j=1}^{n} b_{2j}}{\sum\limits_{i=1}^{n}\sum\limits_{j=1}^{n} b_{ij}} \\[2ex] \cdots \\[1ex] \dfrac{\sum\limits_{j=1}^{n} b_{nj}}{\sum\limits_{i=1}^{n}\sum\limits_{j=1}^{n} b_{ij}} \end{bmatrix} \tag{8.41}$$

The matrix **D** is an estimate of the probability of occurrence for the cause captured in AHP analysis.

However, according to the cause−effect diagram, the faults are not isolated. The probability of a main cause of fault occurring should be estimated from known relationships. This can be done with fault tree analysis (FTA). FTA is a graphical representation related to a particular anomaly of a product, a process or a system. FTA is a top down type analysis in which each of the events that contributes to a particular anomaly can be evaluated in both quantitative and qualitative terms. Once completed, the fault tree allows engineer to fully evaluate the system's safety and reliability by altering the various low-level attributes of the tree.

FTA uses many symbols to represent different elements of the fault tree. Fig. 8.15 shows the FTA symbols that will be used in this analysis.

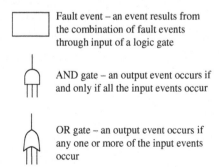

FIGURE 8.15   Some FTA symbols.

In the operational error fault tree, the probabilities of failures can be calculated based on probability logic. An OR gate logic is computed by Eq. (8.42). An AND gate logic is computed by Eq. (8.43) if the events are linked.

$$Pr(D_1 \cup D_2) = Pr(D_1) + Pr(D_2) - Pr(D_1 \cap D_2) \qquad (8.42)$$

$$Pr(D_1 \cap D_2) = Pr(D_1) \times Pr(D_2) \qquad (8.43)$$

If the causes D_1 and D_2 are not linked, $D_1 \cap D_2$ is an empty set, then Pr $(D_1 \cap D_2) = Pr(\Phi) = 0$. Eq. (8.42) becomes

$$Pr(D_1 \cup D_2) = Pr(D_1) + Pr(D_2) \qquad (8.44)$$

Similarly, the operational error or any secondary or higher level faults can be calculated. Based on the prioritized faults, an action plan can then be developed and the resources required to implement the action plan can be budgeted. It should be noted that the operational error has unity probability meaning that if any of the causes of failures occur, the operational error is certain. This property can be used to validate any computational error in the tree. This methodology is illustrated by the following example.

A manufacturing company estimates the total cost of manufacturing errors is $2,949,680 including 2638 h of labor for repair. The errors are categorized into three failure modes, as presented in Table 8.16.

The high level failure modes are then analyzed and regrouped using the cause−effect diagram. Fig. 8.16 shows 22 causes.

The process of FMEA formulation is based on qualitative consideration in point of mangement review for cost effective measures. The FMEA

TABLE 8.16 Cost of Manufacturing Errors According to Failure Modes

Operational Error	Failure in Work Place	Failure in Machine Tooling	Material Failure	Total Defects Cost in Assembly
Drilling error	849,508	424,754	141,585	1,415,847
Assembly error	530,942	265,471	88,490	884,903
Part inspection failure	318,565	159,283	53,094	530,942
Engineering error	117,987	–	–	117,987
Total nonconformance value	1,817,002	849,508	283,169	2,949,679
Percent	61.60%	28.80%	9.60%	100.00%

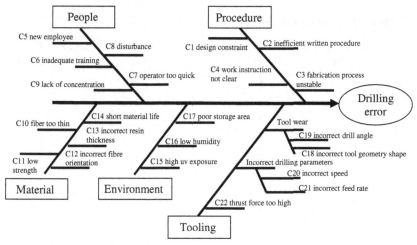

FIGURE 8.16   Cause—effect diagram.

anlaysis classifies drilling errors defects quantiatively by assigning values to factors: severity, occurrences, and detection. The product of these three factors is the RPN which can impact internal failure cost in organization by the bottom—up approach. In this instance, RPN numbers above 150 are considered in the prioritized list in Table 8.17.

Eleven high-priority causes are considered from Table 8.17. These are analyzed by the AHP method, as presented in Table 8.18. The VP has been calculated from AHP and is treated as the probability of events leading to that cause of error.

Based on the paired matrix comparison the value distributed to the three main components for cost-based studies for each cause of failure. The fault tree is then constructed to incorporate the primary causes and the secondary faults, as shown in Fig. 8.17. The probabilities of failures in the tree are computed according to Eqs. (8.42) and (8.43).

The operational risk of drilling errors represents the inability of the manufacturing assembly process to handle the combination of the critical causes of failure in the drilling process.

The overall cost estimation for the prioritized modes of failure can be estimated by combining the probabilities of failures with the recorded cost of manufacturing errors, as presented in Table 8.19. The risk costs are computed by multiplying the categorized costs with the probabilities of failures. The actions to be taken will be designed for the identified causes.

On the basis of Table 8.19, the risk action priority can be used as benchmark to reduce the consequences toward low level to avoid catastrophic. In order to justify the risk avoidance, risk acceptance, and procedural control, risk planning can be assigned to the respective risk owner on operational

TABLE 8.17 Ranked Causes of Drilling Errors

Cause No.	Severity (S)	Occurrences (O)	Detection (D)	Risk Priority Number (RPN)
C6	7	4	7	196
C7	7	4	7	196
C8	7	4	7	196
C9	7	4	7	196
C4	7	5	5	175
C18	5	5	7	175
C19	5	5	7	175
C20	5	5	7	175
C21	5	5	7	175
C22	5	5	7	175
C13	5	5	7	175
C5	7	5	3	105
C15	5	6	3	90
C3	7	4	3	84
C8	5	3	5	75
C7	5	2	7	70
C14	5	2	7	70
C17	5	4	3	60
C2	5	2	5	50
C11	5	3	3	45
C12	5	3	3	45
C13	5	2	3	30
C2	7	3	1	21
C10	5	3	1	15
C10	7	2	1	14
C10	5	2	1	10
C1	5	2	1	10
C12	5	1	1	5

TABLE 8.18 Cause of Failures Prioritized by AHP

	C6	C7	C8	C9	C4	C18	C19	C20	C21	C22	C13	VP
C6	1	1/9	1/3	1/3	1/9	1/3	1/3	1/7	1/9	1/7	1/3	0.01231
C7	9	1	1/9	1/9	1/5	1/7	1/5	1/9	1/9	1/7	1/3	0.02227
C8	3	9	1	1/7	1/3	1/3	1/3	1/3	1/3	1/3	1/3	0.03152
C9	3	9	7	1	9	3	5	1/5	1/3	1/7	7	0.15047
C4	9	5	3	1/9	1	1/9	1/9	1/9	1/9	1/9	1/5	0.03647
C18	3	7	3	1/3	9	1	7	7	7	5	1/7	0.15579
C19	3	5	3	1/5	9	1/7	1	9	7	7	7	0.15869
C20	7	9	3	5	9	1/7	1/9	1	7	5	1/5	0.11691
C21	9	9	3	3	9	1/7	1/7	1/7	1	9	1/5	0.09886
C22	7	7	3	7	9	1/5	1/7	1/5	1/9	1	1/3	0.08867
C13	3	3	3	1/7	5	7	1/7	5	5	3	1	0.12803

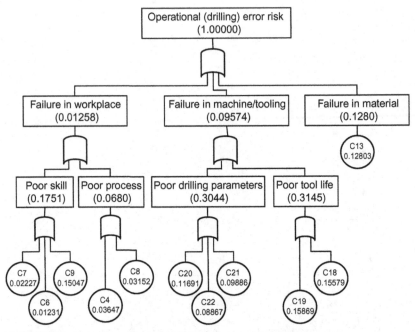

FIGURE 8.17 Fault tree analysis of drilling errors operational risk.

TABLE 8.19 Risk Costs

Faults	Prob.	Categorized Error Costs	Risk Costs	Causes to Take Actions
Drilling failure in machine tooling	0.61023	424,754	259,198	C20, C21, C22
Drilling failure in workplace	0.27524	849,508	233,815	C4, C6, C7, C8, C9
Drilling failure in material	0.11453	141,585	16,216	C13

level to take action. For example, to mitigate the top fault "drilling failure in machine tooling" which can be generated by causes C20, C21, and C22, remedial actions can be taken specific to these causes. Causes C20 "incorrect speed" and C21 "incorrect feed rate" can be avoided by better operation instructions. Cause C22 "Thrust force too high" can be removed by an active force control mechanism to be used in conjunction with the drill. Similarly, decisions on actions to mitigate other causes with lower priorities can be made wherever applicable.

Chapter 9

Beam Deflection

9.1 INTRODUCTION

Beam structures are widely used in many engineering fields, for example, mechanical, civil, and aerospace engineering areas. Many structural components can be approximated as beam elements, a few examples being aircraft wings, automotive drive shafts, and diving boards. Analytical solutions based on solid mechanics theory exist for many types of beam problems. However, there are several complex beam type structures for which an exact solution often is difficult to find, and hence numerical solutions for such cases can serve as a good alternative approach.

9.2 BEAM DEFLECTION PROBLEM FORMULATION

According to the beam theory, a beam is a structural member whose one dimension (length) is significantly larger than the other dimensions (width and height). There are a number of beam theories with the most common being Euler–Bernoulli and Timoshenko beam theories. Here we focus on the simple one—Euler–Bernoulli assumptions, which are the planer cross-section of a beam remains plane after the deformation, and the cross-section remains normal to the neutral axis of the beam [1–3].

Fig. 9.1 shows schematically the configuration of a beam bending with typical parameters shown. Under the Euler–Bernoulli assumptions, Eq. (9.1) [3] gives the curvature of a beam using the notation of Fig. 9.1A [4].

$$\frac{1}{R} = \frac{d^2w/dx^2}{(1+(dw/dx)^2)^{3/2}} \qquad (9.1)$$

Using the assumptions and neglecting the first derivative, Eq. (9.1) becomes Eq. (9.2).

$$\frac{1}{R} = \frac{d^2w}{dx^2} \qquad (9.2)$$

The expression for horizontal strain, in the x direction, is given by Eq. (9.3) [2]. Based on the assumptions, the normal stress is determined using the Hooke's law to be directly proportional to the normal strain and is given in Eq. (9.4).

Demystifying Numerical Models. DOI: https://doi.org/10.1016/B978-0-08-100975-8.00009-6

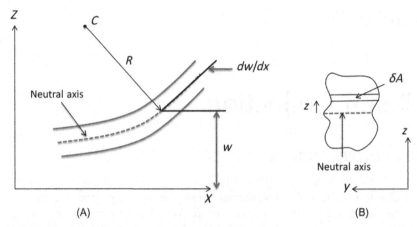

FIGURE 9.1 Schematic of a beam bending with parameters.

$$\varepsilon_x = -\frac{z}{R} \tag{9.3}$$

$$\sigma_x = -E\frac{z}{R} \tag{9.4}$$

This stress generates a force appearing on section δA of the beam, Fig. 9.1B, which then produces a local moment δM (Eq. (9.5)). This could be integrated over the cross-section to find the bending moment M, as shown in Eq. (9.6).

$$\delta M = E\frac{z^2}{R}\delta A \tag{9.5}$$

$$M = \int_A E\frac{z^2}{R}dA \tag{9.6}$$

For the case of a beam with constant material Young's modulus and radius of curvature along the length (i.e., with E and R being constant) and using the definition of the second moment of area of the beam cross-section, I ($I = \int z^2 dA$) the bending moment, M, could be expressed by Eq. (9.7) [3].

$$M = \frac{EI}{R} \tag{9.7}$$

Now, the bending moment can be expressed as a second-order ordinary differential equation of the deflection (Eq. (9.8)) by combining Eqs. (9.7) and (9.2).

$$\frac{M}{EI} = \frac{d^2w}{dx^2} \tag{9.8}$$

For cases with a distributed loading on the beam, we also use another second-order ordinary differential equation (Eq. (9.9)) relating the bending moment M to the distributed load $q(x)$ [4]. This differential equation along with appropriate boundary conditions depending on the loads and constraints on the beam is solved using numerical methods to derive the deformation pattern of beams under bending.

$$\frac{d^2M}{dx^2} = -q(x) \qquad (9.9)$$

9.3 FINITE DIFFERENCE DISCRETIZATION

Eq. (9.8) needs to be discretized to be able to use a numerical method. The method used is the finite difference method (FDM), implemented in a spreadsheet program. The first step is to discretize a beam using a number of points *(N* points), known as grid points. The deflection will be calculated at these points using FDM. Fig. 9.2 shows an example of space discretization with 10 grid points $(N = 10)$.

9.3.1 Interior Points

One part of the finite difference discretization is concerned with interior points (middle points). Boundary conditions will be discussed later. An approximation of a Taylor series is utilized for the discretization. Three types of finite difference schemes can be used for the numerical solution—forward difference, central difference, and backward difference. In each case, the values of the function around a specified point are used to determinate the value of the derivative of the function at that point, as shown in Fig. 9.3.

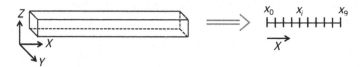

FIGURE 9.2   Example of a discretization with 10 points.

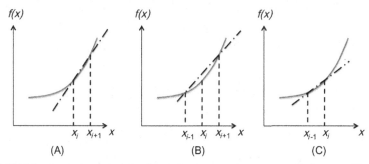

FIGURE 9.3   Approximation of the first derivative at a given point—(A) forward difference, (B) central difference, and (C) backward difference.

TABLE 9.1 Discretization of the First and Second Derivative of a Function Using Taylor's Series

	Forward Difference	Central Difference	Backward Difference
$\frac{\partial f}{\partial x}\big)_{x_i}$	$\frac{f_{x_{i+1}} - f_{x_i}}{\Delta x}$	$\frac{f_{x_{i+1}} - f_{x_{i-1}}}{2\Delta x}$	$\frac{f_{x_i} - f_{x_{i-1}}}{\Delta x}$
$\frac{\partial^2 f}{\partial x^2}\big)_{x_i}$	$\frac{f_{x_i} - 2f_{x_{i+1}} + f_{x_{i+2}}}{\Delta x^2}$	$\frac{f_{x_{i-1}} - 2f_{x_i} + f_{x_{i+1}}}{\Delta x^2}$	$\frac{f_{x_{i-2}} - 2f_{x_{i-1}} + f_{x_i}}{\Delta x^2}$

Derivatives of different orders could be determined. Table 9.1 presents the expressions for first and second derivative of a function using those three finite difference methods.

Using discretization, Eq. (9.8) can be written for each point using central difference for the displacements (w) at ($i-1$)th, ith, and ($i + 1$)th points with coordinate x_i.

$$\frac{w_{i+1} - 2w_i + w_{i-1}}{\Delta x^2} = \frac{M}{EI} \tag{9.10}$$

Eq. (9.10) needs to be solved for each point, and hence a matrix method is used to solve the set of algebraic equations obtained from finite difference discretization. Eq. (9.11) will be used to solve the problem.

$$\mathbf{KU} = \mathbf{V} \tag{9.11}$$

where $\mathbf{U}$ is a vector of the deflection values at N points used in the discretization (w_i); $\mathbf{V}$ is a vector containing right-hand terms given by Eq. (9.8). Thus the values of the interior rows of $\mathbf{V}$ are equal to $M.\Delta x^2/EI$ here. Finally, $\mathbf{K}$ is an $N \times N$ matrix derived from the expression of Taylor series. This gives us Eq. (9.12).

$$\begin{bmatrix} \cdots & \cdots & \cdots & \cdots & \cdots \\ 1 & -2 & 1 & 0 & \cdots \\ 0 & 1 & -2 & 1 & 0 \\ \cdots & 0 & 1 & -2 & 1 \\ \cdots & \cdots & \cdots & \cdots & \cdots \end{bmatrix} \begin{bmatrix} \cdots \\ w_{i-1} \\ w_i \\ w_{i+1} \\ \cdots \end{bmatrix} = \begin{bmatrix} \frac{\ddot{\Delta x^2}M}{EI} \\ \frac{\Delta x^2 M}{EI} \\ \frac{\Delta x^2 M}{EI} \\ \cdots \end{bmatrix} \tag{9.12}$$

Eq. (9.12) can be solved by inverting K matrix to obtain the deflections (U) of different points on the beam using Eq. (9.13).

$$\mathbf{U} = \mathbf{K}^{-1}\mathbf{V} \tag{9.13}$$

Similar approach could be adopted for the forward or backward difference scheme.

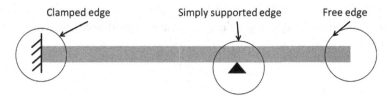

Clamped edge   Simply supported edge   Free edge

FIGURE 9.4 Different types of boundary conditions in a beam.

9.3.2 Boundary Conditions

Each type of beam deflection problem is distinguished by its boundary condition. In beam deformation mechanics, several boundary conditions can be imposed based on the loads and structural connections at various locations of a beam, for example, clamped (fixed), pin joints (simply supported), and roller boundary conditions. Here we will deal with three types of beam deflection problems, a clamped beam, a simply supported beam, and a free edge beam. These different boundary conditions are schematically represented in Fig. 9.4 and are explained below.

9.3.2.1 Clamped Beam

A clamped boundary condition represents a rigidly fixed section/edge of a beam. This implies that the displacement (i.e., deflection) of the given edge is restricted and the tangent to the beam centerline remains horizontal at that point. The corresponding point needs to satisfy Eqs. (9.14) and (9.15) (assuming the clamped section is at $x = x_0$).

$$w\big|_{x_{clamped}} = 0 \tag{9.14}$$

$$\frac{\partial w}{\partial x}\bigg|_{x_{clamped}} = 0$$
$$\Rightarrow w(x_1) - w(x_0) = 0$$
$$\Rightarrow w_1 - w_0 = 0 \tag{9.15}$$

Thus, **K** and **V** matrices need to be modified to consider Eqs. (9.14) and (9.15). In this case, the first and second rows of **K** and **V** are changed for a left clamped edge, and the last and second last rows are changed for a right clamped edge. Eq. (9.15) is an example of a left clamped edge.

$$\begin{bmatrix} 1 & 0 & ... \\ -1 & 1 & ... \\ ... & ... & ... \end{bmatrix} \begin{bmatrix} w_0 \\ w_1 \\ ... \end{bmatrix} = \begin{bmatrix} 0 \\ 0 \\ ... \end{bmatrix} \tag{9.16}$$

9.3.2.2 Simply Supported Beam

A simply supported condition physically represents a pin joint in two dimensional or a ball joint in three dimensional. For a simply supported beam, the deflection is restricted, but the rotation is enabled at the support. For a beam with a simply supported edge, the point at the support needs to satisfy Eqs. (9.17) and (9.18).

$$w\big|_{x_{supported}} = 0 \tag{9.17}$$

$$\frac{\partial^2 w}{\partial x^2}\bigg|_{x_{supported}} = 0$$

$$\Rightarrow w(x_2) - 2w(x_1) + w(x_0) = 0$$

$$\Rightarrow w_2 - 2w_1 + w_0 = 0 \tag{9.18}$$

where $w(x_0) = w_0$, $w(x_1) = w_1$, and $w(x_2) = w_2$

Matrices $\mathbf{K}$ and $\mathbf{V}$ also need to be modified for these equations. Like the clamped beam case, two rows of $\mathbf{K}$ and $\mathbf{V}$ are changed to reflect Eq. (9.18), depending on where the boundary condition is applied. Eq. (9.19) shows modifications for a left simply supported edge.

$$\begin{bmatrix} 1 & 0 & 0 & ... \\ 1 & -2 & 1 & ... \\ ... & ... & ... & ... \\ ... & ... & ... & ... \end{bmatrix} \begin{bmatrix} w_0 \\ w_1 \\ w_2 \\ ... \end{bmatrix} = \begin{bmatrix} 0 \\ 0 \\ ... \\ ... \end{bmatrix} \tag{9.19}$$

9.3.2.3 Free Edge Beam

For this condition, Eq. (9.8) still needs to be solved on considered points. To do so, $\mathbf{K}$ matrix is changed using backward difference for a right free edge and forward difference for a left free edge to set right-hand side of Eq. (9.8) to zero, as a free edge represents zero bending moment, i.e., $M = 0$.

9.4 BEAM DEFLECTION PROBLEMS—NUMERICAL SOLUTIONS

We will demonstrate the solution of beam problems using the Finite Difference Method (FDM) programmed in a spreadsheet application (Microsoft Excel) for different loading and constraints, signifying different

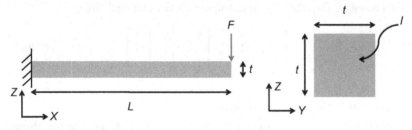

FIGURE 9.5 Schematic of a clamped beam under a point load.

boundary conditions. For all of the beam configurations described in the following, the coordinate system is shown in Fig. 9.5, whereby the origin is located at the left end of the beam with the X-axis along the length of the beam, the Y-axis pointing into the $X-Z$ plane, and the Z-axis pointing vertically upwards. The deflection at any section of a beam is measured along the Z-axis. For each of the beam problem, we use multiple finite difference models, which can be distinguished by the level of discretizations, i.e., the resolution of grid points chosen. A number of beam deflection problems have been solved using the finite difference method implemented in a spreadsheet program, and the results are presented below [4].

9.4.1 Clamped Beam Under a Point Load

The first case considered is to find the deflection of a beam, which is clamped at one (left) edge and free at the other edge with a point force (F) acting at the free edge, as shown in Fig. 9.5. The beam has a square cross-section with side t. For this example case, the beam parameter values have been taken as—length $L = 1$ m, the point force $F = 100$ N, and the width and height $t = 59.5$ mm $= 0.0595$ m. With these parameter values, the second moment of area becomes $I = 1.042 \times 10^{-6}$ m^4. The beam is a made of aluminum with a Young's modulus $E = 70$ GPa and a Poisson's ratio $\nu = 0.35$.

For this problem, the bending moment $M(x)$ at any section x is given by Eq. (9.20) [2].

$$M(x) = (L - x)F \qquad (9.20)$$

Using the forward difference equation and the clamped boundary condition to modify **K** matrix and using Eq. (9.20) to construct **V** matrix, it is required to calculate numerically the deflection of the beam along the length. To study the convergence of the deflection solution, four finite difference (FD) models of the beam were developed, each with different number of grid points (resolutions). The four models had increasing number of grid points, 6, 20, 40, 100 points along the length of the beam. Fig. 9.6 summarizes the results from the various FD models with different levels of discretization. The deflection profile of the beam (along the Z-axis) along the length (along the X-axis) of the beam is shown in Fig. 9.6. It can be noted that as the number of points increases, the deflection profile apporaches gradually to a specific converged solution, with little diference between the deflection profile obtained with 40 and 100 points.

9.4.2 Simply Supported Beam Under a Point Load

Next we consider a simply supported beam under a point load. This problem is to find the deflection of the beam that is simply supported on each edge, and is acted on by a point load at a distance a from the left end (and b from

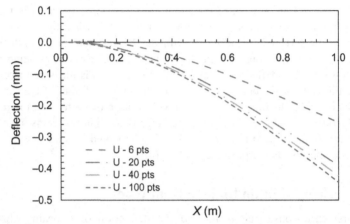

FIGURE 9.6 Deflection profile of a clamped beam with a point load along its length, showing the effect of number of points on the deflection solutions.

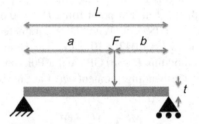

FIGURE 9.7 Schematic of a simply supported beam under a point load.

the right end), as shown in Fig. 9.7. The length of the beam is L $(a + b = L)$ and it is of square cross-section of side t. The FD models are created for parameters $L = 1$ m, $a = 0.7$ m, $b = 0.3$ m, $t = 0.0595$ m, $I = 1.042 \times 10^{-6}$ m^4, and $F = 1000$ N. As before, the beam is considered to be made of aluminum with $E = 70$ GPa and $\nu = 0.35$.

The bending moment $M(x)$ can be expressed by Eq. (9.21) [2].

$$x \in [0, a] \rightarrow M(x) = \frac{F}{EI}\left(1 - \frac{a}{L}\right)x$$

$$x \in [a, L] \rightarrow M(x) = \frac{F}{EI}(L - x)\frac{a}{L}$$

(9.21)

For finite difference discretization, we use the central difference method along with simply supported boundary conditions, and the matrices **K** and **V** are amended accordingly. In this case, two finite difference models with discretization of 10 and 20 points were used because the results are converged and agreed well with the analytical solution. Fig. 9.8 shows the deflection

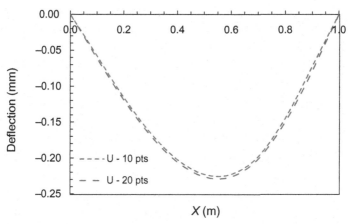

FIGURE 9.8 Deflection profiles of a simple supported beam under a point load acting at 70% of its length from the left edge for two different discretizations of 10 and 20 points along the beam length.

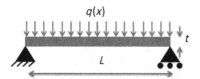

FIGURE 9.9 Schematic of a beam under a uniformly distributed load $q(x)$.

profiles of the beam along the length for 10 and 20 points discretizations. The resulting profiles are close, with the maximum deflection occurring at the point of application of the force.

9.4.3 Simply Supported Beam Under a Uniform Load

Next the case of a beam with uniformly distributed load is considered, as shown in Fig. 9.9. The geometry and material properties of the beam are the same as used previously, i.e., $L = 1$ m, $t = 59.5$ mm, $I = 1.042 \times 10^{-6}$ m^4, $E = 70$ GPa, and $\nu = 0.35$. The solution for a constant $q(x) = q = 1000$ N/m is presented in Fig. 9.10. In this problem, the bending moment $M(x)$ at a section x is related to the load using Eq. (9.22) [3].

$$\frac{d^2M}{dx^2} = -q \tag{9.22}$$

Integrating twice and using the boundary conditions, $M(0) = M(L) = 0$, $M(x)$ can be expressed using Eq. (9.23) [2].

$$M(x) = \frac{qx}{2}(L - x) \tag{9.23}$$

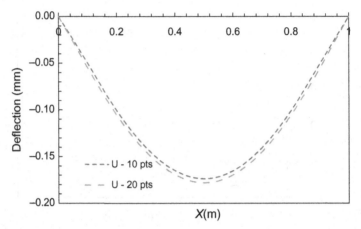

FIGURE 9.10 Deflection profiles of a uniformly loaded beam for two different discretizations of 10 and 20 points along the beam length.

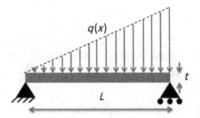

FIGURE 9.11 Schematic of a simply supported beam under a linear load.

Using Eqs. (9.23) and (9.8), **V** matrix can be formulated. In this case also, two finite difference models with discretizations of 10 and 20 points were used because the results converged and agreed well with analytical equation. The results for the deflections along the beam length are shown in Fig. 9.10. The maximum deflection is 1.78 mm and occurs at the middle section of the beam.

9.4.4 Simply Supported Beam Under a Linear Load

Following the case of a uniform load, we consider a more complex load case of a linearly varying load along the length, as shown in Fig. 9.11. The maximum value of the load intensity at the right end is $q = 10$ N/m with the linear distribution given by Eq. (9.24), which provides the load intensity at a distance x from the left end of the beam [2].

$$q(x) = \frac{qx}{L} \tag{9.24}$$

The geometry and material properties of the beam are the same as before, i.e., $L = 1$ m, $t = 59.5$ mm, $I = 1.042 \times 10^{-6}$ m^4, $E = 70$ GPa, and $\nu = 0.35$. Using the new distribution of $q(x)$, $M(x)$ is now given by Eq. (9.25) [2].

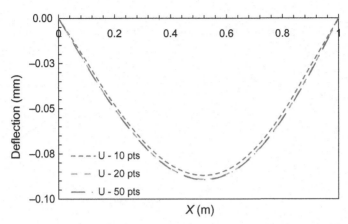

FIGURE 9.12 Deflection profiles of a linearly distributed loaded beam for three different discretizations of 10, 20, and 50 points along the beam length.

$$M(x) = \frac{qLx}{6}\left(1 - \frac{x^2}{L^2}\right) \tag{9.25}$$

Simply supported boundary conditions at the left and right ends are used, and the matrices **K** and **V** are constructed accordingly. The solution for maximum $q(x) = 1000$ N/m is presented in Fig. 9.12. For this beam configuration, the beam deflection geometry obtained from three FD models with 10, 20, and 50 points are close, with the solutions from the latter two models are almost the same. So the solutions converged and a discretization of 20 points is sufficient for this problem to obtain a nearly grid point independent result.

9.4.5 Functionally Graded Clamped Beam Under a Point Load With a Linearly Varying Elastic Modulus

The aim of this section is to demonstrate that it is feasible to solve numerically relatively complex beam problems that may be difficult to solve analytically or may not even have an exact solution. To demonstrate this, we choose a functionally graded beam with varying material stiffness or elastic modulus (Young's modulus) along the length. The geometry and boundary condition of the beam are the same as before (Section 9.4.1), i.e., $L = 1$ m, $t = 59.5$ mm, and $I = 1.042 \times 10^{-6}$ m^4. However, the Young's modulus ranges from $E_2 = 30$ GPa at the fixed edge (left end) to $E_1 = 110$ GPa at the free edge. These values have been taken to ensure a mean Young's modulus equal to 70 GPa, which is the value used for the constant material property beams. The analytical solution for such problems is usually complex, and it is often effective to solve these problems using numerical methods. We have used same equations, Eqs. (9.8) to (9.12) to solve the problem, but each row

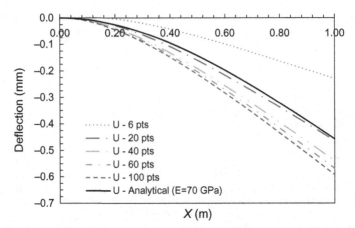

FIGURE 9.13 Deflection profiles for the functionally graded beam with different discretizations and their comparison with the analytical solution for a beam with constant Young's modulus.

of **V** matrix is now different because of E variation with the ith row of Eq. (9.12) modified as in Eq. (9.26).

$$V_i = \frac{\Delta x^2 M}{E_i I} \qquad (9.26)$$

where E_i is the Young's modulus of the beam at the ith grid point location. Fig. 9.13 shows the results of this problem for various levels of discretizations (6, 20, 40, 60, and 100 points), and the solution (the deflection profile) for this relatively complex problem is found to converge. The results obtained from 40, 60, and 100 points are quite close. The analytical solution of this problem assuming constant Young's modulus is also compared. It can be seen that the deflection profile (brown dotted line) of the functionally graded beam is highly different from that of the analytical solution (black solid line) with constant E. A functionally graded beam with the same mean Young's modulus has a much larger deflection, compared with constant property beam.

9.5 CONVERGENCE AND ACCURACY

In numerical methods, convergence and accuracy are of prime considerations. In the FDMs, the numerical solution should approach a specific value with increasing number of grid points used to discretize the problem. This feature is known as the convergence of a solution. The converged solution should be ideally almost independent of the number of grid points. The second aspect is accuracy of a solution, which can be assessed by verification or validation of a numerical solution against an analytical (exact) solution or experimental result. For the beam problems studied, we now evaluate the accuracy of the finite difference models by comparing the numerical solutions with the analytical (exact) solutions.

We compare the beam deflection profiles from various finite difference models with different grid resolutions with the analytical solutions of the deflected beam profile for all of the beam configurations considered, as characterized by loading and boundary conditions.

The error in the numerical method is defined as the relative error of the maximum deflection as follows:

$$Error = \frac{w_{num} - w_{ana}}{w_{ana}} \qquad (9.27)$$

where w_{num} and w_{ana} are the numerical and analytical solutions, respectively. For instance, for the clamped beam with a point load at the free end, the analytical solution is given by Eq. (9.28) [3]

$$w = \frac{-Fx^2}{6EI}(3L - x) \qquad (9.28)$$

As expected, Fig. 9.14 shows that with an insufficient number of points, the numerically obtained deflection deviates significantly from the exact solution. The deflection results match closely with an increasing number of points, i.e., for a high-resolution model.

The expected predicted maximum deflection value, as given by the analytical solution, is $w_{max} = -0.457$ mm. The errors with progressive finer grid resolutions (i.e., the number of points) are presented in Table 9.2. As the model grid resolution increases, the solution (the maximum deflection)

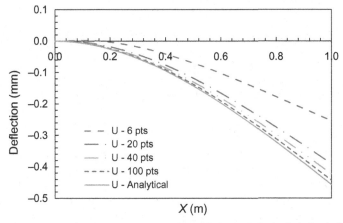

FIGURE 9.14 Comparison between the numerical solution and the analytical solution for a clamped beam with a point load at the free edge.

TABLE 9.2 Comparison of the Maximum Deflections Predicted by the Numerical Models With the Analytical Solution for a Clamped Beam With a Point Load at the Free Edge

Number of Points	Maximum Deflection (mm)	Error (%)
6	−0.254	44.4
20	0.391	14.5
40	0.423	7.4
100	−0.444	3.0

approaches the analytical solution. The error reduces rapidly as the number of points is increased. The error for 6 points is quite high (44.4%) and reduced to only 3% when the model used 100 points.

We have compared the numerical solutions obtained for different grid resolutions with the corresponding analytical solution for the other three cases, namely simply supported beam with a point load, with a uniformly distributed load, and linearly distributed load.

This analytical solutions for these cases is presented in Eqs. (9.29), (9.30), and (9.31) [3,4].

For simply supported beam with a point load:

$$x \in [0, a] \to w(x) = \frac{Fb}{6LEI}(x^3 + a(a - 2L)x)$$

$$x \in [a, L] \to w(x) = \frac{-Fa}{6LEI}(x^3 - 3Lx^2 + (a^2 + 2L^2)x - a^2L)$$

(9.29)

For simply supported beam with uniformly distributed load:

$$w(x) = \frac{q}{24EI}(2Lx^3 - x^4 - L^3x)$$

(9.30)

For simply supported beam with linearly distributed load:

$$w(x) = \frac{q}{360EIL}(-3x^5 + 10L^2x^3 - 7L^4x)$$

(9.31)

For all of these beam configurations, the beam deflection profiles predicted numerically are compared with the corresponding analytical solutions in Figs. 9.15−9.17. It can be seen that with a grid resolution of 20 points, the numerical solutions are quite accurate in capturing the exact deflected shapes, as given by the analytical expressions in Eqs. (9.29) and (9.30). Even for a complex linearly distributed load case, the solutions for 20 and 50 points are indistinguishable from the analytical solution.

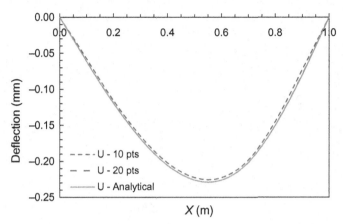

FIGURE 9.15 Comparison between the numerical solutions and the analytical solution for a simply supported beam with a point load.

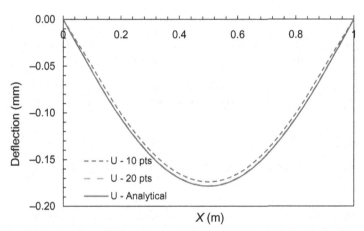

FIGURE 9.16 Comparison between the numerical solutions and the analytical solution for a simply supported beam with a uniformly distributed load.

When evaluating the quality of the numerical solutions in terms of errors, it is important to analyse the error in the most significant parameter of interest, i.e., the maximum deflection value in this case, and also the errors elsewhere, i.e., the maximum error in the deflection value along the length of a beam. Table 9.3 presents the former errors for two resolutions of 10 and 20 points with respect to the analytical value of $w_{max} = 0.229$ mm, which is compared with the two numerical solutions.

Table 9.4 summarizes the maximum errors for the simply supported beam with different loading conditions and different grid resolutions. The

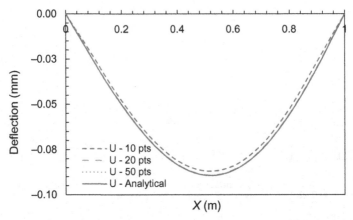

FIGURE 9.17 Comparison between the numerical solutions and the analytical solution for a simply supported beam with a linearly distributed load.

TABLE 9.3 Comparison Between the Numerical Solutions With Two Grid Resolutions With the Analytical Solution for the Maximum Deflection of a Simply Supported Beam With a Point Load at the Free Edge

Number of Points	Numerical Maximum Deflection (mm)	Error %
10	− 0.223	1.5
20	− 0.229	0.02

TABLE 9.4 Maximum Errors in Deflection Relative to the Analytical Solution for a Simply Supported Beam for Different Loads and Grid Resolutions

Beam Configuration (Simply Supported)	Number of Points	Maximum Error (%)
Point load	10	9.3
	20	2.4
Uniformly distributed load	10	10.09
	20	2.63
Linearly distributed load	10	11.48
	20	3.12
	50	0.53

maximum error usually occurs at locations around which boundary conditions are applied. It is noteworthy that in the case of point loading, the error in maximum deflection is very small, 0.02% for 20 points, whereas the maximum error is 2.4%. Based on the maximum errors, finite difference models with about 20 points are sufficiently accurate with maximum errors being in the range of 2%–3%. The error can be significantly reduced for a higher resolution model. For example, for the linearly distributed load case, the maximum error is $\sim 0.5\%$ with 50 points, producing a highly accurate solution.

9.6 SUMMARY

This chapter deals with an important category of structural component, i.e., the beam elements, which are widely used in numerous structural applications, e.g., aerospace, transport, marine, mining, and machinery. It has been demonstrated that the FDM is an effective tool for accurate prediction of deformation of beam structures under different loads and structural constraints. The method can be implemented in a spreadsheet program (e.g., Microsoft Excel) to solve for beam problems by applying suitable boundary conditions. Not only the solutions converge to the respective analytical solution with a reasonable grid point resolution, but also relatively complex beam configurations, whereby an exact solution does not exist or difficult to obtain, can be solved by simple use of the FDM in a spreadsheet program.

REFERENCES

[1] Megson THG. Chapter 3 - Normal force, shear force, bending moment and torsion. Structural and stress analysis. 3rd ed. Boston: Butterworth-Heinemann; 2014. p. 38−78.

[2] Megson THG. Chapter 9 - Bending of beams. Structural and stress analysis. 3rd ed. Boston: Butterworth-Heinemann; 2014. p. 209−52.

[3] Megson THG. Chapter 13 - Deflection of beams. Structural and stress analysis. 3rd ed. Boston: Butterworth-Heinemann; 2014. p. 337−88.

[4] Application of Numerical Methods for the Solution of Solid Mechanics Problems. Pantaloni, D., and Das, R., 2017. RMIT University, Australia, Report No. 2017/2A-SIM.

Chapter 10

Mechanical Vibration

10.1 INTRODUCTION TO VIBRATION PROBLEMS

Vibration problems are encountered in many engineering applications. A range of vibration problems are regularly encountered in mechanical, civil, and aerospace engineering areas. Whenever a component or structure moves or is subjected to a periodic motion, it is termed as vibration. The effect of vibration and resonance can generate extreme loads or deformations of a structural component, leading to catastrophic failures in some cases. As a result, vibrational analysis has become an integral part of mechanical, structural, and acoustic designs. For example, aircraft wings are required to be designed to eliminate any possibility of resonance, which could lead to severe consequences during in-flight conditions. Critical civil structures, such as bridges and tall buildings, need to be designed taking wind-induced vibration into consideration. Collapse of Tacoma Narrows Bridge in 1940 caused by wind resonance is an example of severity of vibration effects on critical engineering structures.

10.2 VIBRATION PROBLEM FORMULATION

We will consider here one-dimensional (1D) vibration for simplicity. The body is considered as a point mass to develop the equation of motion and vibration. For a real three-dimensional body, the mass of the body is assumed to be concentrated as a point mass at the center of mass (G). The Newton's second law for 1D motion of a mass provides the equation of motion in (Eq. 10.1):

$$m\ddot{u} = \sum F_{ext} \tag{10.1}$$

where u and m are the displacement and mass of the body, and F_{ext} is an external force acting on the body. Vibrations are caused by restoring forces and are opposed by resistive or damping forces, such as friction force. The expressions for the restoring and resistive forces could be expressed as given in Eq. (10.2), where k and c are constant and u is the displacement of the body (a point mass or its center of mass for a distributed mass) from its equilibrium position [1].

Demystifying Numerical Models. DOI: https://doi.org/10.1016/B978-0-08-100975-8.00010-2
209

$$F_r = -ku$$
$$F_d = -c\dot{u}$$ $\qquad$ (10.2)

where k and c are constant and u is the displacement of the body from its equilibrium position. It is possible to express the displacement x as $x = x_0 + u$, where x_0 is the original location of the mass. Considering a body subjected to a restoring force, a friction force and a set of external forces independent of u, the equation of motion in Eq. (10.1) could be modified by substituting F_r and F_d from Eq. (10.2) to provide Eq. (10.3) [1].

$$m\ddot{u} = F_r + F_d + \sum F_{ext}$$
$$\ddot{u} + \frac{c}{m}\dot{u} + \frac{k}{m}u = \frac{1}{m}\sum F_{ext}$$ $\qquad$ (10.3)

Eq. (10.3) could also be written as Eq. (10.4) using parameters $\delta = c/2m$ and $\omega_0 = \sqrt{k/m}$, where ω_0 is the natural frequency and δ is the damping factor,

$$\ddot{u} + 2\delta\dot{u} + \omega_0^2 u = \frac{1}{m}\sum F_{ext}$$ $\qquad$ (10.4)

Eq. (10.4) is the differential equation that will be used later.

10.3   FINITE DIFFERENCE DISCRETIZATION

Vibration problems mainly deal with temporal or time derivatives, rather than spatial derivatives. Hence the ordinary differential equation represented by Eqn. (10.4) needs to be discretized with respect to time, rather than space. The first step is to discretize the total range of time using a number of points (N points), known as time steps, at which the values of displacement u are estimated (Fig. 10.1).

10.3.1   Interior Points

The finite difference discretization has a number of interior points (middle points). After the time range is discretized, a numerical formulation of Eq. (10.4) using the backward difference method can be developed. Initial conditions are represented by variable values at $t = 0$, i.e., the displacement and velocity at the start are known as the initial condition, and no information on u is known after this time. Hence for this type of problems, the backward finite difference scheme is a suitable method. The discretized form of Eq. (10.4) is expressed as Eq. (10.5).

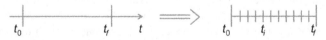

FIGURE 10.1   Schematic of a finite difference discretization with respect to time.

$$\frac{u_i - 2u_{i-1} + u_{i-2}}{\Delta t^2} + 2\delta \frac{u_i - u_{i-1}}{\Delta t} + \omega_0^2 u_i = \frac{1}{m}\sum F_{ext} \qquad (10.5)$$

where Δt is the time step.

Eq. (10.5) needs to be solved at each time step, and hence Eq. (10.5) is then expanded in a matrix form in Eq. (10.6), which incorporates the initial conditions. A matrix method is then used to solve the set of algebraic equations obtained from the finite difference discretization as in Eq. (10.6).

$$\left(\frac{1}{\Delta t^2}K_1 + \frac{2\delta}{\Delta t}K_2 + \omega_0^2 K_3\right)U = V \qquad (10.6)$$

The matrices K_1, K_2, and K_3 can be expressed in Eq. (10.7). These matrices can be included in a spreadsheet program and can be solved.

$$\left(\frac{1}{\Delta t^2}\begin{bmatrix} \dots & \dots & \dots & \dots & \dots \\ 1 & -2 & 1 & 0 & \dots \\ 0 & 1 & -2 & 1 & 0 \\ \dots & 0 & 1 & -2 & 1 \\ \dots & \dots & \dots & \dots & \dots \end{bmatrix} + \frac{2\delta}{\Delta t}\begin{bmatrix} \dots & \dots & \dots & \dots & \dots \\ -1 & 1 & 0 & 0 & \dots \\ 0 & -1 & 1 & 0 & 0 \\ \dots & 0 & -1 & 1 & 0 \\ \dots & \dots & \dots & \dots & \dots \end{bmatrix} + \omega_0^2\begin{bmatrix} \dots & \dots & \dots & \dots & \dots \\ 0 & 1 & 0 & 0 & \dots \\ 0 & 0 & 1 & 0 & 0 \\ \dots & 0 & 0 & 1 & 0 \\ \dots & \dots & \dots & \dots & \dots \end{bmatrix}\right)\begin{bmatrix} \dots \\ u_{i-2} \\ u_{i-1} \\ u_i \end{bmatrix} = \frac{1}{m}\begin{bmatrix} \dots \\ \sum F_{ext} \\ \sum F_{ext} \\ \sum F_{ext} \end{bmatrix}$$

$$(10.7)$$

Solution of Eq. (10.7) by matrix inversion function in the spreadsheet, we can directly obtain the displacement of the object on each discretized time using the initial conditions.

10.3.2 Initial Conditions

Vibrations are time-dependent (transient or temporal) problems, with often only the initial conditions are specified. Here the most common initial conditions, a specified displacement and a velocity at time $t = 0$, are used. These initial conditions can be expressed using Taylor series in a discretized manner. The specified displacement ($u_{initial}$) and velocity ($\dot{u}_{initial}$) at $t = 0$ are given in Eq. (10.8).

$$\begin{aligned} u_0 &= u_{initial} \\ u_1 - u_0 &= \dot{u}_{initial}\Delta t \end{aligned} \qquad (10.8)$$

These two relationships need to be inserted in Eq. (10.7). These are inserted in matrix K_1 as shown in Eqn. 10.7. This leads to modification of matrices V, K_2 and K_3 with their first two rows set equal to zero and U remaining unchanged.

$$
\mathbf{K}_1 = \begin{bmatrix} \Delta t^2 & 0 & \dots & \dots & \dots \\ -\Delta t & \Delta t & 0 & \dots & \dots \\ 1 & -2 & 1 & 0 & \dots \\ 0 & 1 & -2 & 1 & 0 \\ & & \dots & & \end{bmatrix} \quad \text{and} \quad \mathbf{V} = \frac{1}{m} \begin{bmatrix} mu_{initial} \\ m\dot{u}_{initial} \\ \sum F_{ext} \\ \sum F_{ext} \\ \dots \end{bmatrix} \qquad (10.9)
$$

All matrices in Eqn.10.7 are now known, and hence the displacement vector $\mathbf{U}$ can be determined.

10.4   VIBRATION PROBLEMS—NUMERICAL SOLUTIONS

We will illustrate the use of a spreadsheet program (such as Microsoft Excel) for solving various categories of vibration problems. There are primarily two kinds of vibrations, free and forced vibrations. A free vibration is a periodic motion of an object with no external forces on it, whereas a forced vibration involves a periodic motion under externally impressed forces on the object. The damping in a vibration (known as damped vibration) is the resistance to the motion of an object, for example, due to dry friction or fluid friction. The damping is expressed by matrix $\mathbf{K}_2$, thus for vibration without damping, $\mathbf{K}_2 = \mathbf{0}$. A number of vibration problems have been solved using the finite difference method implemented in a spreadsheet program, and the results are presented below [1,2].

10.4.1   Free Vibration Without Damping

1D free vibration is the simplest type of vibration. This problem is typically an object attached to a spring, as shown in Fig. 10.2. For this case, the natural frequency (ω_0) is expressed by Eq. (10.10).

$$
\omega_0 = \sqrt{\frac{k}{m}} \qquad (10.10)
$$

where k is the spring stiffness and m is the mass of the object. For the example problem considered, the time range is taken to be 100 s. The problem is solved for varying parameters, natural frequency (ω_0), initial displacement

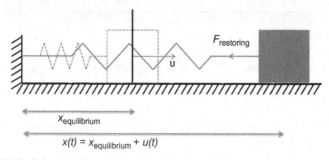

FIGURE 10.2   Schematic of an object subjected to free vibration.

(u_0), and initial velocity ($\dot{u}_0$), to study the effects of these parameters on the displacement, see Figs. 10.3−10.6. Each model was solved for various discretizations levels (10, 50, 100, and 500 points). It can be seen that sufficient points or time steps are required to obtain an accurate periodic response. For example, the solution with 10 points is highly inaccurate and is unable to capture the periodic response.

The results demonstrate the effects of the different parameters. Both the initial velocity and initial displacement affect the amplitude and phase of vibration, and the natural frequency influences the time period (i.e., frequency) of vibration. As it is shown later, these effects also hold true even for forced vibration. It is to be noted from Fig. 10.5 numerical errors caused

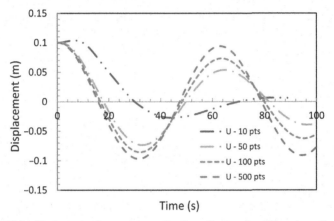

FIGURE 10.3 Displacement-time response of a free vibration for different number of time steps (discretized points in time) with $\omega_0 = 0.1 \text{ s}^{-1}$, $u_0 = 0.1$ m, and $\dot{u}_0 = 0$ m s^{-1}.

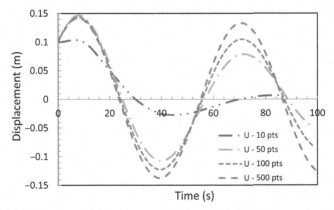

FIGURE 10.4 Displacement-time response of a free vibration for different number of time steps (discretized points in time) with $\omega_0 = 0.1 \text{ s}^{-1}$, $u_0 = 0.1$ m, and $\dot{u}_0 = 0.01$ m s^{-1}.

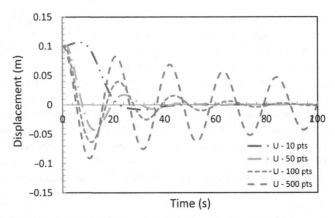

FIGURE 10.5 Displacement-time response of a free vibration for different number of time steps (discretized points in time) with $\omega_0 = 0.3 \text{ s}^{-1}$, $u_0 = 0.1$ m, and $\dot{u}_0 = 0 \text{ m s}^{-1}$.

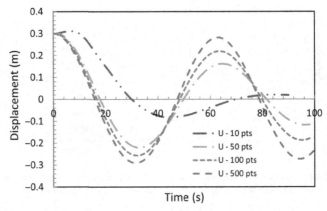

FIGURE 10.6 Displacement-time response of a free vibration without damping for different number of time steps (discretized points in time) with $\omega_0 = 0.1 \text{ s}^{-1}$, $u_0 = 0.3$ m, and $\dot{u}_0 = 0 \text{ m s}^{-1}$.

by insufficient discretization may lead to an artificial damping type effect in the system.

10.4.2   Forced Vibration Without Damping

The configuration of a forced vibration is similar to the previous one of free vibration except in addition a periodic external force (F_{ext}) is applied on the object, whereby $F_{ext} = F \sin(\omega t)$ (Fig. 10.7). For this example, the time range is again taken to be 100 s and an external periodic force of amplitude 0.01 N ($F = 0.01$ N) is applied. The vibration parameters are set equal to: $\omega_0 = 0.1 \text{ s}^{-1}$, $m = 1$ kg, $k = 0.01 \text{ N m}^{-1}$, $u_0 = 0.1$ m, and $\dot{u}_0 = 0 \text{ m s}^{-1}$.

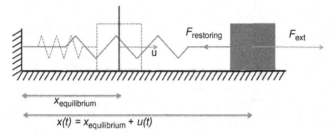

FIGURE 10.7 Schematic of an object subjected to forced vibration.

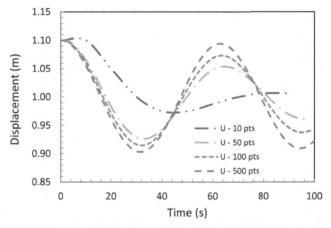

FIGURE 10.8 Displacement-time response of a forced vibration without damping for different number of time steps (discretized points in time) with $F = 0.01$ N, $\omega_0 = 0.1$ s^{-1}, $u_0 = 0.1$ m, and $\dot{u}_0 = 0$ m s^{-1}.

Fig. 10.8 shows the displacement of the object with time for different levels of discretization, incorporating 10, 50, 100, and 500 points (i.e., time steps) over the time. It is to note that a sufficient number of points or time steps are required to improve the accuracy of the solution.

The characteristics of forced vibrations are similar in many respects to those of free vibrations. An offset in displacement is observed depending on the spring stiffness (k) and the amplitude of the force (F), as given in Eq. (10.11). This offset value is equal to the equilibrium position of the corresponding static configuration of the system.

$$offset = \frac{F}{m\omega_0^2} = \frac{F}{k} \tag{10.11}$$

10.4.3 Free Vibration With Damping

Next we consider free vibration in the presence of damping (Fig. 10.9). The key parameter that affects the vibration is the damping factor (δ). Depending

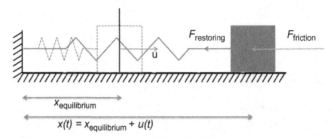

FIGURE 10.9 Schematic of a free vibration with damping.

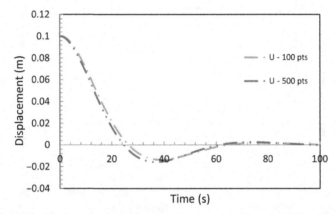

FIGURE 10.10 Displacement-time response of a free underdamped vibration for different number of time steps (discretized points in time) with $\omega_0 = 0.1 \text{ s}^{-1}$, $\delta = 0.05 \text{ s}^{-1}$, $u_0 = 0.1 \text{ m}$, and $\dot{u}_0 = 0 \text{ m s}^{-1}$.

on the relative values of damping factor and natural frequency (ω_0), there are three cases, underdamped, critically damped, and overdamped systems. The time range is taken to be 100 s for all of the cases shown below. Each case has been solved repeatedly with several discretizations in time, usually 100, and 500 points or time steps.

Case (i): Underdamping ($\delta < \omega_0$)

For the underdamped case, the object oscillates over several cycles with diminishing amplitude due to damping present and finally the motion is drastically reduced with very low displacement. For this case, different parameters, including initial displacement (u_0) and velocity ($\dot{u}_0$), natural frequency (ω_0), and most importantly damping ratio (δ), were varied to study the effects of each one on the vibration, as shown in Figs. 10.10–10.12.

It was found that the initial velocity has a crucial effect on the displacement, in terms of changing the amplitude and the phase. The damping factor has a crucial effect. The closer the damping factor is to the natural frequency,

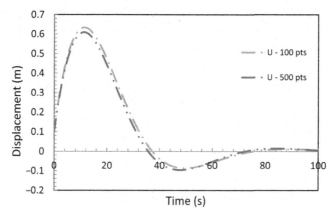

FIGURE 10.11 Displacement-time response of a free underdamped vibration for different number of time steps (discretized points in time) with $\omega_0 = 0.1 \text{ s}^{-1}$, $\delta = 0.05 \text{ s}^{-1}$, $u_0 = 0.1$ m, and $\dot{u}_0 = 0.1 \text{ m s}^{-1}$.

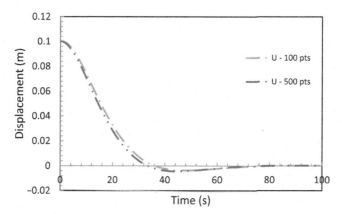

FIGURE 10.12 Displacement-time response of a free underdamped vibration for different number of time steps (discretized points in time) with $\omega_0 = 0.1 \text{ s}^{-1}$, $\delta = 0.07 \text{ s}^{-1}$, $u_0 = 0.1$ m, and $\dot{u}_0 = 0 \text{ m s}^{-1}$.

the faster the object reaches close to its equilibrium position. The effects of initial displacement and natural frequency are same in that they modify the amplitude of vibrations and the time period, as seen before.

Case (ii): Critical Damping ($\delta = \omega_0$)

For a critically damped system, the vibratory motion terminates when the object reaches the equilibrium position, i.e., when for the first time $u = 0$. Fig. 10.13 shows a critically damped system with zero initial velocity, and Fig. 10.14 shows one with nonzero initial velocity ($\dot{u}_0 = 0.2 \text{ m.s}^{-1}$). These

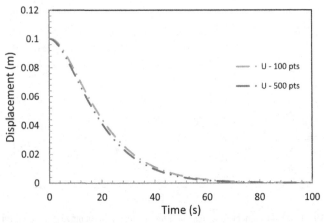

FIGURE 10.13 Displacement-time response of a free critically damped vibration for different number of time steps (discretized points in time) with $\omega_0 = 0.1 \text{ s}^{-1}$, $\delta = 0.1 \text{ s}^{-1}$, $u_0 = 0.1$ m, and $\dot{u}_0 = 0 \text{ m s}^{-1}$.

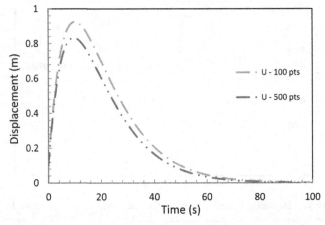

FIGURE 10.14 Displacement-time response of a free critically damped vibration for different number of time steps (discretized points in time) with $\omega_0 = 0.1 \text{ s}^{-1}$, $\delta = 0.1 \text{ s}^{-1}$, $u_0 = 0.1$ m, and $\dot{u}_0 = 0.2 \text{ m s}^{-1}$.

two cases shown are solved for $\delta = \omega_0 = 0.1 \text{ s}^{-1}$. It is noted that the finite difference scheme needs at least 100 points or time steps to reach a good quality solution that predicts the nature of the motion.

Case (iii): Overdamping ($\delta > \omega_0$)

Next we consider an over damping condition in vibration. In theoretical, the object attains its equilibrium after an infinite time. As previously, Figs. 10.15 and 10.16 demonstrate the influence of damping ratio on the

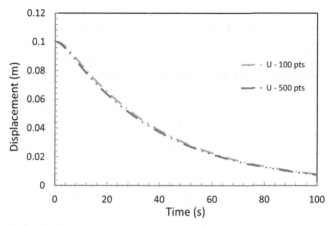

FIGURE 10.15 Displacement-time response of a free overdamped vibration for different number of time steps (discretized points in time) with $\omega_0 = 0.1 \text{ s}^{-1}$, $\delta = 0.2 \text{ s}^{-1}$, $u_0 = 0.1$ m, and $\dot{u}_0 = 0 \text{ m s}^{-1}$.

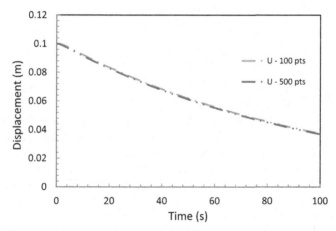

FIGURE 10.16 Displacement-time response of a free overdamped vibration for different number of time steps (discretized points in time) with $\omega_0 = 0.1 \text{ s}^{-1}$, $\delta = 0.5 \text{ s}^{-1}$, $u_0 = 0.1$ m, and $\dot{u}_0 = 0 \text{ m s}^{-1}$.

vibration response. It is noteworthy that a small number of time steps (e.g., 50 points) is not adequate to capture the vibration response. The use of sufficient discretization, such as 100 points or more, provides similar displacement response.

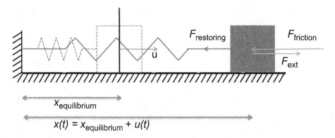

FIGURE 10.17 Schematic of a forced vibration with damping.

Figs. 10.15 and 10.16 show the effect of the damping factor on the vibration response. The higher the damping factor is, the longer the time required to reach closer to the equilibrium position.

10.4.4 Forced Vibration With Damping

The same problem as before is considered with a periodic applied force to render the case of forced vibration with damping, which has three cases, namely underdamping, critical damping, and overdamping (Fig. 10.17). The static equilibrium is now given by Eq. (10.12).

$$u_{equilibrium} = \frac{F}{m\omega_0^2} = \frac{F}{k} \tag{10.12}$$

For all of the following cases, ω_0 and m are selected in a way so as to always produce $k = 0.005 \text{ N m}^{-1}$. Furthermore, the external force is $F = 0.01 \text{ N}$. Hence the static equilibrium position is the same in all cases, and it is given by $u_{equilibrium} = 2 \text{ m}$. It is noteworthy that the effects of all factors stated in the previous section 10.4.3 remain valid. The three cases of vibration are solved using spreadsheet with two different time resolutions with the number of points or time steps being 100 and 500, respectively.

Case (i): Underdamping ($\delta < \omega_0$)

The nature of the vibrations is same as free vibration; however, the motion now tends to reach towards the equilibrium position, $u_{equilibrium}$ as given by Eqn. 10.12. Fig. 10.18 shows an underdamped vibration with the initial displacement being equal to 0.1 m.

As in free damped vibration, the displacement crosses the equilibrium position and oscillates around it.

Case (ii): Critical Damping ($\delta = \omega_0$)

As observed in the free damped vibration, the displacement terminates at the equilibrium position, Fig. 10.19.

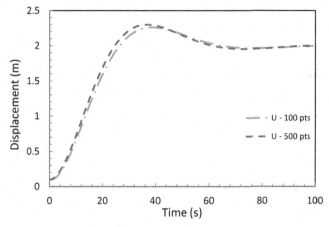

FIGURE 10.18 Displacement-time response of a forced underdamped vibration for different number of time steps (discretized points in time) with $F = 0.01$ N, $\omega_0 = 0.1$ s^{-1}, $\delta = 0.05$ s^{-1}, $u_0 = 0.1$ m, and $\dot{u}_0 = 0$ m s^{-1}.

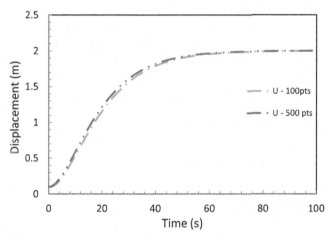

FIGURE 10.19 Displacement-time response of a forced critically damped vibration for different number of time steps (discretized points in time) with $F = 0.01$ N, $\delta = \omega_0 = 0.1$ s^{-1}, $u_0 = 0.1$ m, and $\dot{u}_0 = 0$ m s^{-1}.

Case (iii): Overdamping $(\delta > \omega_0)$

For the overdamped case solved using two different time resolutions (number of points or time steps), it is expected that the displacement will never reach or pass the equilibrium Fig. 10.20.

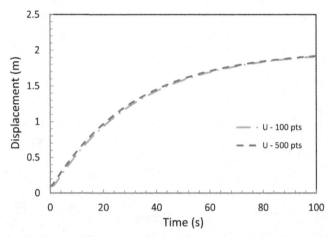

FIGURE 10.20 Displacement-time response of a forced overdamped vibration for different number of time steps (discretized points in time) with $F = 0.01$ N, $\omega_0 = 0.25$ s^{-1}, $\delta = 1$ s^{-1}, $u_0 = 0.1$ m, and $\dot{u}_0 = 0$ m s^{-1}.

10.5 CONVERGENCE

For vibration problems, the error can be defined in a slightly different way to avoid large error values caused by a denominator close to zero. In this case, we use a mean error (*Err*) as given in Eq. (10.13).

$$Err = \frac{|u_{num} - u_{ana}|}{(|u_{num}| + |u_{ana}|)/2} \qquad (10.13)$$

where u_{num} and u_{ana} are the numerical and analytical solutions, respectively.

10.5.1 Free Vibration Without Damping

The problem of free vibration without damping has a simple analytical solution, which is given by Eq. (10.14) [2].

$$u(t) = A\cos(\omega t - \varphi) \qquad (10.14)$$

where A and φ are obtained using initial condition. On this case, a spreadsheet solver (Microsoft Excel) file has been used.

Effect of Natural Frequency on Error

The analytical solution is difficult to approximate for a given time range. Firstly, the sampling frequency ($f_{sampling}$) should meet the Shannon condition Eq. (10.15) to ensure good model estimation.

$$f_{sampling} \geq 2\omega_0 \qquad (10.15)$$

Furthermore, some errors are introduced by numerical methods in each period, which reduces the maximum displacement in each period. This effect is more pronounced when the sampling frequency decreases. The error is more manifested if more cycles are present, i.e., the higher is the frequency, the more important the mean error is, see Figs. 10.21 and 10.22. In Fig. 10.21, a numerical model with 500 points (time steps) predicts the solution very close to the analytical solution for $\omega_0 = 0.1 \text{ s}^{-1}$; however, when the frequency is increased three times to $\omega_0 = 0.3 \text{ s}^{-1}$, the same number of points was not able to predict the maximum displacement (amplitude) well and produced a large error, as seen in Fig. 10.21. Fig. 10.23 shows that the mean error increases drastically with frequency for a given discretization.

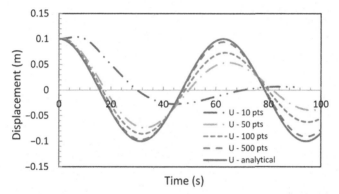

FIGURE 10.21 Comparison between the numerical and analytical solutions for a free vibration without damping and for $\omega_0 = 0.1 \text{ s}^{-1}$, $u_0 = 0.1 \text{ m}$, and $\dot{u}_0 = 0 \text{ m s}^{-1}$.

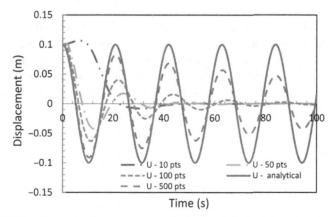

FIGURE 10.22 Comparison between the numerical and analytical solutions for a free vibration without damping and for $\omega_0 = 0.3 \text{ s}^{-1}$, $u_0 = 0.1 \text{ m}$ and $\dot{u}_0 = 0 \text{ m s}^{-1}$.

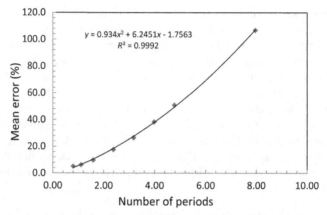

FIGURE 10.23 Relationship between the mean error and number of periods for 500 time steps.

TABLE 10.1 Variation of Mean Error with Natural Frequency for 500 Time Steps

ω_0 (s^{-1})	0.05	0.07	0.1	0.15	0.2	0.25	0.3	0.5
Number of periods	0.80	1.11	1.59	2.39	3.18	3.98	4.77	7.96
Mean error (%)	5.1	6.2	9.8	17.7	26.5	38.3	50.8	106.7

Fig. 10.21 shows the nature of variation of mean error with the number of periods is approximately quadratic. This is primarily because with an increase in ω_0, the number of periods within a time range increases, which requires more time steps to capture the periodic response accurately.

It is seen from Fig. 10.23 that the mean error increases with the number of periods for a specified time range (Table 10.1).

Effect of Initial Velocity On Error

The initial velocity affects the phase (φ) of vibration for a given natural frequency and amplitude. Table 10.2 presents that the mean error decreases as the magnitude of initial velocity increases. Fig. 10.24 shows the variation of the mean error with phase which supports the same finding that with an increase in the phase, the mean error decreases, in this case from nearly 10% to 5%.

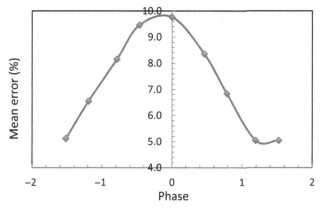

FIGURE 10.24   Effect of initial velocity on mean error for the numerical solution with 500 time steps.

10.5.2   Forced Vibration Without Damping

As stated in Section 10.4.2, this case is close to the case of free vibration without damping, but it has an offset. The analytical or exact solution is given by Eq. (10.16) [1].

$$u(t) = A\cos(\omega t - \varphi) + \frac{F}{k} \tag{10.16}$$

Hence this problem is expected to show similar convergence to that of free vibration without damping. Fig. 10.25 shows the error is large when few time steps (≤ 100) are used. When the number of time steps is increased, the numerical solution approaches the analytical solution very closely, which follows the similar behavior seen for the corresponding free vibration problem in Fig. 10.21.

10.5.3   Free Vibration With Damping

Numerical solutions for the free vibration with damping are compared with the corresponding analytical solutions for the three categories of vibrations, underdamped, critically damped, and overdamped systems.

Case (i): Underdamping ($\delta < \omega_0$)

In this case underdamped vibration is considered for the convergence studies. First free vibration with damping is studied. The analytical solution is given by Eq. (10.17) [1].

TABLE 10.2 Effect of Initial Velocity on Mean Error for the Numerical Solution with 500 Time Steps

Phase	−1.52	−1.19	−0.79	−0.46	0	0.46	0.79	1.19	1.52
Initial velocity (m s⁻¹)	−0.2	−0.025	−0.01	−0.005	0	0.005	0.01	0.025	0.2
Mean error (%)	5.12	6.55	8.15	9.46	9.77	8.36	6.83	5.05	5.05

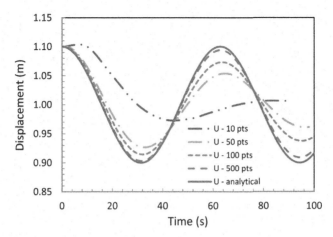

FIGURE 10.25 Comparison between the numerical and analytical solutions for a free vibration without damping and for $F = 0.01$ N, $k = 0.01$ N m^{-1}, $\omega_0 = 0.1$ s^{-1}, $u_0 = 0.1$ m, and $\dot{u}_0 = 0$ m s^{-1}.

$$u(t) = Ae^{-\delta t}\cos\left(\sqrt{\omega_0^2 - \delta^2}\,t - \varphi\right)$$ (10.17)

where A and φ are computed based on initial conditions. The mean error is affected by the damping factor. The mean error with 500 steps with a damping factor (ratio) equal to 0.02 is 13.59%, while it becomes 18.95% with a damping factor equal to 0.05 (Fig. 10.26).

Case (ii): Critical Damping ($\delta = \omega_0$)

For the case of critical damping, the analytical solution is given by Eq. (10.18) [1].

$$u(t) = (A + Bt)e^{-\delta t}$$ (10.18)

where A and B are evaluated based on initial conditions. Fig. 10.27 shows the comparison between the analytical and two numerical solutions with different discretizations of 100 and 500 time steps, whereby the numerical solutions match closely with the analytical one.

The maximum error is affected by the natural frequency and the initial velocity. It should be noted that when the analytical solution is close to zero, the relative error becomes higher.

Case (iii): Overdamping ($\delta > \omega_0$)

The analytical solution for the overdamped system is given by Eq. (10.19) [1]. The maximum error is used here as the solution does not cross the x-axis

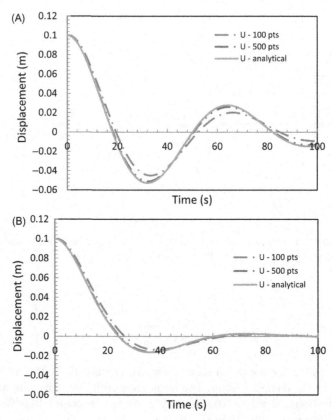

FIGURE 10.26 Comparison between the numerical and analytical solutions for a free under-damped vibration with damping ratio (a) $\delta = 0.02$ s^{-1} and (b) $\delta = 0.05$ s^{-1} (for $\omega_0 = 0.1$ s^{-1}, $u_0 = 0.1$ m and $\dot{u}_0 = 0$ m s^{-1}.

(nonzero solution). Hence the magnitude of error would be less sensitive to the solution.

$$u(t) = \left(u_0 \cosh(\alpha t) + \frac{\delta u_0 + \dot{u}_0}{\alpha} \sinh(\alpha t) t \right) e^{-\delta t}$$

(10.19)

$$\text{with} \quad \alpha = \sqrt{\delta^2 - \omega_0^2}$$

Fig. 10.28 shows the comparison between the two numerical solutions with different discretizations of 100 and 500 time steps and the analytical solution, whereby the numerical solutions match closely with the analytical one. The maximum error is 1.25% for a damping ratio of $\delta = 0.2$ s^{-1}.

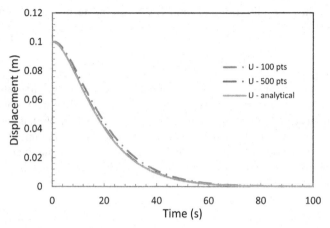

FIGURE 10.27   Comparison between the numerical and analytical solutions for a free critically damped vibration with $\delta = \omega_0 = 0.1 \text{ s}^{-1}$, $u_0 = 0.1$ m and $\dot{u}_0 = 0 \text{ m s}^{-1}$.

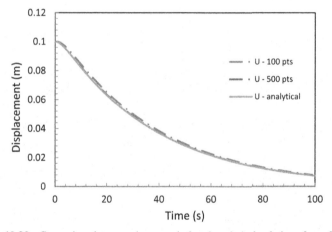

FIGURE 10.28   Comparison between the numerical and analytical solutions for a free over-damped vibration with $\delta = 0.2 \text{ s}^{-1}$, $\omega_0 = 0.1 \text{ s}^{-1}$ $u_0 = 0.1$ m and $\dot{u}_0 = 0 \text{ m s}^{-1}$.

10.5.4   Forced Vibration With Damping

Numerical solutions for the forced vibration with damping are compared with the corresponding analytical solutions for the three categories of vibrations, underdamped, critically damped, and overdamped systems.

Case (i) Underdamping ($\delta < \omega_0$)

This underdamped vibration problem has an analytical solution given by Eq. (10.20) [1]. For the forced vibration, an offset term F/k is added to the free vibration solution for a forcing term with a sinusoidal excitation.

$$u(t) = Ae^{-\delta t}\cos(\sqrt{\omega_0^2 - \delta^2}\, t - \phi) + \frac{F}{k} \tag{10.20}$$

The comparison between the numerical (finite difference) and analytical solutions is shown in Fig. 10.29 for two different initial velocity conditions, which indicates a good agreement between the numerical and analytical solutions. An error analysis shows that with an increase in the initial velocity (i.e., initial condition), the relative accuracy of the numerical solution improves for this case. For the higher initial velocity of $\dot{u}_0 = 1\ \mathrm{m\ s^{-1}}$ (Fig. 10.29B), the maximum error is only 0.87%.

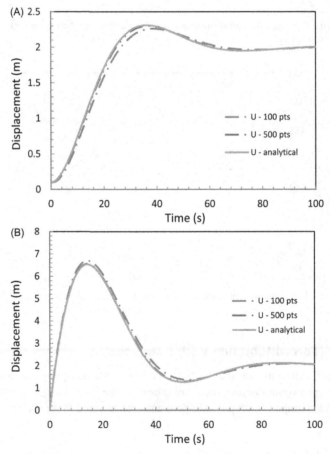

FIGURE 10.29 Comparison between the numerical and analytical solutions for forced under-damped vibration with (a) $\dot{u}_0 = 0\ \mathrm{m\ s^{-1}}$ and (b) $\dot{u}_0 = 1\ \mathrm{m\ s^{-1}}$ ($F = 0.01$ N, $k = 0.005$ N m^{-1}, $\delta = 0.05\ \mathrm{s^{-1}}$, $\omega_0 = 0.1\ \mathrm{s^{-1}}$, and $u_0 = 0.1$ m).

Case (ii): Critical Damping ($\delta = \omega_0$)

This critically damped case has the analytical solution given by Eq. (10.21) [1], and A and B can be calculated by Eq. (10.22).

$$u(t) = (A + Bt)e^{-\delta t} + \frac{F}{k} \qquad (10.21)$$

$$A = u_0 - \frac{F}{k} \quad \text{and} \quad B = \dot{u}_0 + \delta\left(u_0 - \frac{F}{k}\right) \qquad (10.22)$$

Fig. 10.30 shows the comparison of the numerical and analytical solutions for two different initial velocities, zero and nonzero values. The solution agreements seem to be quite good, with the maximum error being 1.35% for the nonzero velocity case with a greater number of time steps.

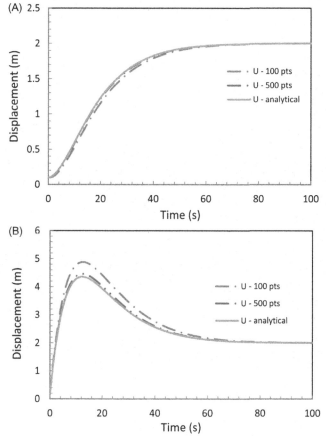

FIGURE 10.30 Comparison between the numerical and analytical solutions for forced critically damped vibration with (a) $\dot{u}_0 = 0 \text{ m s}^{-1}$ and (b) $\dot{u}_0 = 1 \text{ m s}^{-1}$ ($F = 0.01$ N, $k = 0.005$ N m^{-1}, $\delta = \omega_0 = 0.1$ s^{-1}, and $u_0 = 0.1$ m).

Case (iii): Overdamping ($\delta > \omega_0$)

For the overdamped case, the analytical solution is presented in Eq. (10.23) [1] where X_1 and X_2 are dependent on the initial conditions as shown in Eq. (10.24).

$$u(t) = \left(X_1 e^{-\alpha t} + X_2 e^{+\alpha t}\right)e^{-\delta t} + \frac{F}{k} \text{ with } \alpha = \sqrt{\delta^2 - \omega_0^2} \qquad (10.23)$$

$$X_1 = u_0 - X_2 - \frac{F}{k} \text{ and } X_2 = \frac{1}{2\alpha}\left(\dot{u}_0 + (\alpha + \delta)\left(u_0 - \frac{F}{k}\right)\right) \qquad (10.24)$$

Fig. 10.31 shows the comparison between the analytical and numerical solutions to be in agreement for the two numerical solutions with zero and nonzero initial velocities ($\dot{u}_0$). Each solution was obtained using a coarse

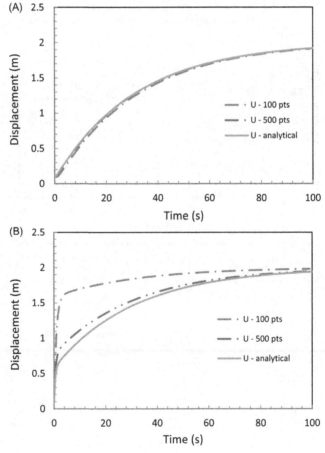

FIGURE 10.31 Comparison between the numerical and analytical solutions for forced over-damped vibration with (a) $\dot{u}_0 = 0$ m s^{-1} and (b) $\dot{u}_0 = 1$ m s^{-1} ($F = 0.01$ N, $k = 0.005$ N m^{-1}, $\delta = 1$ s^{-1}, $\omega_0 = 0.25$ s^{-1}, and $u_0 = 0.1$ m).

and a fine discretization (100 and 500 time steps). The mean error for coarse resolution becomes high with an increase in the initial velocity.

10.6   SUMMARY

Vibration problems are encountered in many engineering applications. The Newton's second law provides the equation of motion with system parameters including displacement, mass of a body, and external forces acting on a body. Vibrations are caused by restoring forces and are opposed by resistive or damping forces, such as friction force. The ordinary differential equation derived from this fundamental principle can then be solved by the finite difference discretization method. In this chapter, several categories of one-dimensional simple vibration problems were solved using the finite difference method. The primary reason for considering this is that the present problems are simple and have analytical solutions. Hence, this type of problems provides a good basis against which numerical solutions obtained using a spreadsheet program (Microsoft Excel) could be verified. Both free and forced vibrations with and without damping were considered. Numerical discretization or the number of time steps was varied to demonstrate its effect on the solution quality. The errors and convergence of the numerical solutions were assessed by comparing them against the analytical solutions. The method explained in this chapter can be generalized to three-dimensional problems using the matrix formulation.

REFERENCES

[1]   Inman DJ. Engineering vibration. 4th ed. USA: Pearson; 2013.
[2]   Application of Numerical Methods for the Solution of Solid Mechanics Problems. Pantaloni, D., and Das, R., 2017. RMIT University, Australia, Report No. 2017/2A-SIM.

Chapter 11

Thin Plate Deflection

11.1 INTRODUCTION

Plate structures are widely used in many engineering fields, for example, aerospace, mechanical, and automotive engineering disciplines. The theory for plate defletion is more complicated than beam deflection. Nevertheless, it is necessary for many engineering areas. As the beam theory is highly important in civil engineering, the plate theory is equally an important part of transport engineering, whereby the use of plate and shell structures is very prevalent, especially in aerospace engineering.

Analytical solutions based on elasticity theory exist for many types of beam as well as plate problems. However there are several complex plate structures for which an exact solution often is difficult to find, and hence numerical solutions for such cases can serve as a good alternative approach. This chapter presents a few plate deformation problems which have known analytical solutions to be able to verify the numerical solutions. Engineers may prefer numerical solutions for complex problems, particularly where the analytical solutions are difficult to obtain.

11.2 PLATE DEFLECTION PROBLEM FORMULATION

Similar to the beam theory, there are some assumptions made for the plate theory. The first assumption is related to the geometry of a structure. A plate is defined to be a structure with two of the dimensions (length and width) considerably larger than the third one (thickness). The second assumption is similar to Euler−Bernoulli assumption. The neutral plane of a plate remains plane after bending. This enables derivation of an analytical relationship between the load on a plate (q) and the deflection (w) [1]. Another main assumption for the problems considered in this chapter is "thin plate theory." A plate is treated as a thin one, when the thickness is considerably smaller than its planar dimensions, with a typical thickness to width ratio being <0.1.

Equilibrium of a plate element is considered to follow the differential equation when the plate is subjected to transverse, bending, and twisting loads. Fig. 11.1 shows a plate element and all those applied conditions.

Demystifying Numerical Models. DOI: https://doi.org/10.1016/B978-0-08-100975-8.00011-4

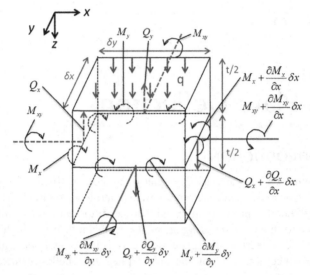

FIGURE 11.1   Schematic of a plate deformation under transverse, bending, and twisting loads.

The moments, M_x, M_y and M_{xy} (Eq. (11.1)), and the vertical shear forces applied per unit width of the plate, Q_x and Q_y (Eq. (11.2)), can be related to the normal stress (σ_{xy}, σ_{yz}, and σ_{zx}) and shear stress (τ_{xy}, τ_{yz}, and τ_{zx}) [2,3].

$$M_x = \int_{-t/2}^{t/2} \sigma_x z \, dz \qquad M_y = \int_{-t/2}^{t/2} \sigma_y z \, dz \qquad M_{xy} = -M_{yx} = -\int_{-t/2}^{t/2} \tau_{xy} z \, dz$$

$$(11.1)$$

$$Q_x = \int_{-t/2}^{t/2} \tau_{xz} \, dz \qquad Q_y = \int_{-t/2}^{t/2} \tau_{yz} \, dz \qquad (11.2)$$

The flexural stiffness, denoted by D, of the plate is expressed by Eq. (11.3) in terms of the thickness (t), Young's modulus (E), and Poisson's ratio (v). The relationships between the moments and the transverse displacement or deflection (w) can be obtained in terms of the flexural stiffness and Poisson's ratio of the plate [2], and these are expressed in Eq. (11.4).

$$D = \frac{Et^3}{12(1 - v^2)} \qquad (11.3)$$

$$M_x = -D\left(\frac{\partial^2 w}{\partial x^2} + v\frac{\partial^2 w}{\partial y^2}\right) \quad M_y = -D\left(\frac{\partial^2 w}{\partial y^2} + v\frac{\partial^2 w}{\partial x^2}\right) \quad M_{xy} = D(1 - v)\frac{\partial^2 w}{\partial x \partial y}$$

$$(11.4)$$

For a distributed load of q, a local static equilibrium condition is applied to obtain Eq. (11.5) for the moment about the x-axis and the twist in the xy plane and Eq. (11.6) for the vertical shear force.

$$M_{xy}\delta y - \left(M_{xy} + \frac{\partial M_{xy}}{\partial x}\delta x \right)\delta y - M_y\delta x + \left(M_y + \frac{\partial M_y}{\partial y}\delta y \right)\delta x$$

$$- \left(Q_y + \frac{\partial Q_y}{\partial y}\delta y \right)\delta x\delta y + Q_x\frac{\delta y^2}{2} - \left(Q_x + \frac{\partial Q_x}{\partial x}\delta x \right)\frac{\delta y^2}{2} - q\delta x\frac{\delta y^2}{2} = 0$$

$$(11.5)$$

$$\left(Q_x + \frac{\partial Q_x}{\partial x}\delta x \right)\delta y - Q_x\delta y + \left(Q_y + \frac{\partial Q_y}{\partial y}\delta y \right)\delta x - Q_y\delta x + q\delta x\delta y = 0 \quad (11.6)$$

After simplifying and ignoring the second-order terms, Eq. (11.5) reduces to Eq. (11.7) and Eq. (11.6) becomes Eq. (11.8). Applying the same about the y-axis and simplifying, Eq. (11.9) is derived.

$$\frac{\partial M_{xy}}{\partial x} - \frac{\partial M_y}{\partial y} + Q_y = 0 \qquad (11.7)$$

$$\frac{\partial Q_x}{\partial x} + \frac{\partial Q_y}{\partial y} + q = 0 \qquad (11.8)$$

$$\frac{\partial M_{xy}}{\partial y} - \frac{\partial M_x}{\partial x} + Q_x = 0 \qquad (11.9)$$

Substituting Q_x from Eq. (11.9) and Q_y from Eq. (11.8) in Eq. (11.7), Eq. (11.10) is obtained.

$$\frac{\partial^2 M_x}{\partial x^2} - 2\frac{\partial^2 M_{xy}}{\partial x \partial y} + \frac{\partial^2 M_y}{\partial y^2} = -q \qquad (11.10)$$

Replacing M_x, M_y, and M_{xy} in Eq. (11.10) from Eq. (11.4), Eq. (11.11) is obtained, which is the differential equation that will be solved numerically in this chapter using a spreadsheet solver.

$$\frac{\partial^4 w}{\partial x^4} + 2\frac{\partial^4 w}{\partial x^2 \partial y^2} + \frac{\partial^4 w}{\partial y^4} = \frac{q}{D} \qquad (11.11)$$

11.3 FINITE DIFFERENCE DISCRETIZATION

Eq. (11.11) needs to be discretized to be able to use a numerical method. The method used is the finite difference method (FDM) in the two-dimensional plane over the plate geometry, which was implemented in a

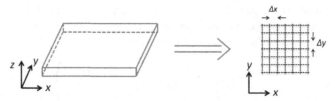

FIGURE 11.2   Schematic of a two-dimensional discretization of a plate.

spreadsheet program. The first step is to discretize a plate using a number of points in the plane of the plate. The spacings between the adjacent points are Δx and Δy, with n and m grid points in the x and y directions, respectively, as shown in Fig. 11.2. It is assumed that $\Delta x = \Delta y = \Delta$ to simplify the discretization. The deflection (the out of plane displacement) of the plate at a grid point (x_i, y_j) (with $x_i = (i\text{-}1)^*\Delta$ and $x_j = (j\text{-}1)^*\Delta$) is denoted by $w_{i,j}$.

11.3.1 Interior Points

As discussed in Chapter 9, Beam Deflection, there are three methods of discretization, forward, backward, and central difference methods. In each case, the values of the deflection around a specified point are used to determinate the high-order derivatives of deflection at that point. This problem uses the central difference method, although forward and backward difference methods can also be adopted. Eq. (11.11) is discretized using the Taylor series. The discretization of Eq. (11.11) for each of the terms is presented in Eqs. (11.12)–(11.14). The detailed steps of the derivation are given in [2,3].

$$\frac{\partial^4 w_{i,j}}{\partial x^4} = \frac{w_{i+2,j} - 4w_{i+1,j} + 6w_{i,j} - 4w_{i-1,j} + w_{i-2,j}}{\Delta^4} \tag{11.12}$$

$$\frac{\partial^4 w_{i,j}}{\partial y^4} = \frac{w_{i,j+2} - 4w_{i,j+1} + 6w_{i,j} - 4w_{i,j-1} + w_{i,j-2}}{\Delta^4} \tag{11.13}$$

$$\frac{\partial^4 w_{i,j}}{\partial x^2 \partial y^2} = \frac{\begin{array}{c}4w_{i,j} - 2(w_{i,j-1} + w_{i,j+1} + w_{i-1,j} + w_{i+1,j}) \\ + w_{i+1,j-1} + w_{i-1,j-1} + w_{i+1,j+1} + w_{i-1,j+1}\end{array}}{\Delta^4} \tag{11.14}$$

Based on Eqs. (11.12)–(11.14), the detailed steps of the derivation are given in [2,3] and subsequent simplification, the discretized forms of Eq. (11.11) for the interior points are obtained as an algebraic equation, given by Eq. (11.15).

$$\frac{1}{\Delta^4}\left(\begin{array}{c}20w_{i,j} - 8(w_{i,j-1} + w_{i,j+1} + w_{i-1,j} + w_{i+1,j}) \\ + 2(w_{i+1,j-1} + w_{i-1,j-1} + w_{i+1,j+1} + w_{i-1,j+1}) \\ + w_{i,j-2} + w_{i,j+2} + w_{i-2,j} + w_{i+2,j}\end{array}\right) = \frac{q}{D} \tag{11.15}$$

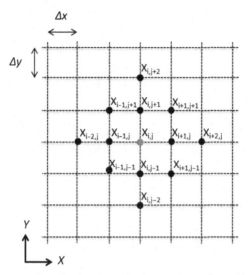

FIGURE 11.3 Representation of an interior grid point and surrounding grid points.

It is to note that Eq. (11.15) expresses the deflection at a grid point (x_i, y_j) in terms of the deflections at the neighboring grid points both in the x and y directions. This equation can be only applied in the interior points, whereby coordinates of all the grid points are known, as seen in Fig. 11.3.

Eq. (11.15) needs to be solved for each point. Similar to the beam deflection problem, a matrix method is used to solve the set of simultaneous algebraic equations obtained from the finite difference discretization.

Eq. (11.16) will be used to solve the problem.

$$\mathbf{KU} = \mathbf{V} \qquad (11.16)$$

However unlike the similar equation in Chapter 9, Beam Deflection, these matrices ($\mathbf{K}$, $\mathbf{U}$, and $\mathbf{V}$) are different for this problem. $\mathbf{U}$ is a matrix of $n \times m$ values of deflections at the grid points, where n and m are the number of grid points along the x-axis and y-axis, respectively. $\mathbf{U}$ is represented by Eq. (11.17).

$$\mathbf{U} = \begin{bmatrix} \mathbf{U}_1 \\ \dots \\ \mathbf{U}_i \\ \dots \\ \mathbf{U}_n \end{bmatrix} \text{ with } \mathbf{U}_i = \begin{bmatrix} w_{i,1} \\ \dots \\ w_{i,j} \\ \dots \\ w_{i,m} \end{bmatrix} \qquad (11.17)$$

$\mathbf{V}$ is a matrix of dimension $n \times m$ that represents the right-hand term of Eq. (11.15), which is expressed by Eq. (11.18).

$$V = \begin{bmatrix} \dfrac{q}{D} \\ \cdots \\ \dfrac{q}{D} \\ \cdots \\ \dfrac{q}{D} \end{bmatrix} \qquad (11.18)$$

K is a $n \times m$ by $n \times m$ matrix with its elements representing the Taylor series coefficients (for the interior points) of Eq. (11.15). It is a block diagonal represented by Eq. (11.19) as shown.

$$K = \begin{bmatrix} A & B & C & 0 & 0 & 0 & 0 \\ B & A & B & C & 0 & 0 & 0 \\ C & B & A & B & C & 0 & 0 \\ 0 & C & B & A & B & C & 0 \\ 0 & 0 & C & B & A & B & C \\ 0 & 0 & 0 & C & B & A & B \\ 0 & 0 & 0 & 0 & C & B & A \end{bmatrix} \text{ with}$$

$$A = \begin{bmatrix} 20 & -8 & 1 & 0 & 0 & 0 & 0 \\ -8 & 20 & -8 & 1 & 0 & 0 & 0 \\ 1 & -8 & 20 & -8 & 1 & 0 & 0 \\ 0 & 1 & -8 & 20 & -8 & 1 & 0 \\ 0 & 0 & 1 & -8 & 20 & -8 & 1 \\ 0 & 0 & 0 & 1 & -8 & 20 & -8 \\ 0 & 0 & 0 & 0 & 1 & -8 & 20 \end{bmatrix}$$

$$(11.19)$$

$$B = \begin{bmatrix} -8 & 2 & 0 & 0 & 0 & 0 & 0 \\ 2 & -8 & 2 & 0 & 0 & 0 & 0 \\ 0 & 2 & -8 & 2 & 0 & 0 & 0 \\ 0 & 0 & 2 & -8 & 2 & 0 & 0 \\ 0 & 0 & 0 & 2 & -8 & 2 & 0 \\ 0 & 0 & 0 & 0 & 2 & -8 & 2 \\ 0 & 0 & 0 & 0 & 0 & 2 & -8 \end{bmatrix} \text{ and}$$

$$C = \begin{bmatrix} 1 & 0 & 0 & 0 & 0 & 0 & 0 \\ 0 & 1 & 0 & 0 & 0 & 0 & 0 \\ 0 & 0 & 1 & 0 & 0 & 0 & 0 \\ 0 & 0 & 0 & 1 & 0 & 0 & 0 \\ 0 & 0 & 0 & 0 & 1 & 0 & 0 \\ 0 & 0 & 0 & 0 & 0 & 1 & 0 \\ 0 & 0 & 0 & 0 & 0 & 0 & 1 \end{bmatrix}$$

Eq. (11.16) can be solved by inverting $\mathbf{K}$ matrix to obtain the deflections ($\mathbf{U}$) of different points on the plate.

$$\mathbf{U} = \mathbf{K}^{-1}\mathbf{V} \tag{11.20}$$

This provides the deflection of a plate under a known load. A similar approach could be adopted for a forward or backward difference scheme.

11.3.2 Boundary Conditions

Two types of boundary conditions are considered for the four edges of a plate, simply supported and clamped edges.

11.3.2.1 Simply Supported Edges

Similar to a beam, a simply supported plate satisfies the conditions that both the deflection and moments are zero at the simply supported edge. The zero moment leads to second derivative of the deflection to be zero as well. These boundary conditions are represented by Eqs. (11.21) and (11.22) for all points at the simply supported edge, i.e., $x = x_{\text{supported}}$.

$$w\big|_{x_{\text{supported}}} = 0 \tag{11.21}$$

$$\frac{\partial^2 w}{\partial x^2}\bigg|_{x_{\text{supported}}} = 0 \Rightarrow w(x_2) - 2w(x_1) + w(x_0) = 0 \tag{11.22}$$

For a simply supported edge at a fixed y, x is replaced by y in Eqs. (11.21) and (11.22). Matrices $\mathbf{K}$ and $\mathbf{V}$ need to be modified to incorporate these equations stipulated by the boundary conditions. The nature of the modification depends on the alignment of the coordinate axis with the edge. If the simple supported edge is along the x-axis, $\mathbf{K}$ and $\mathbf{V}$ matrices are modified according to Eqs. (11.23) and (11.24).

$$\mathbf{K} = \begin{bmatrix} \mathbf{A}_{\text{lim1}} & \mathbf{B}_{\text{lim1}} & \mathbf{C}_{\text{lim1}} & \mathbf{0} & \dots \\ \mathbf{B}'_{\text{lim2}} & \mathbf{A}_{\text{lim2}} & \mathbf{B}_{\text{lim2}} & \mathbf{C}_{\text{lim2}} & \mathbf{0} \\ \mathbf{C} & \mathbf{B} & \mathbf{A} & \mathbf{B} & \mathbf{C} \\ \dots & \dots & \dots & \dots & \dots \end{bmatrix} \tag{11.23}$$

with $\mathbf{A}_{\text{lim1}} = \mathbf{B}'_{\text{lim2}} = \mathbf{B}_{\text{lim2}} = \mathbf{I}_m$ and $\mathbf{A}_{\text{lim2}} = -2\mathbf{I}_m$ and $\mathbf{C}_{\text{lim1}} = \mathbf{C}_{\text{lim2}} = \mathbf{0}$

$$\mathbf{V} = \begin{bmatrix} \mathbf{V}_1 \\ \dots \\ \mathbf{V}_i \\ \dots \\ \mathbf{V}_n \end{bmatrix} \text{ with } \mathbf{V}_i = \begin{bmatrix} \dfrac{q}{D} \\ \dots \\ \dfrac{q}{D} \\ \dots \\ \dfrac{q}{D} \end{bmatrix} \text{ and } \mathbf{V}_{1,2} = \begin{bmatrix} 0 \\ \dots \\ \dots \\ \dots \\ 0 \end{bmatrix} \tag{11.24}$$

If the alignment of the simply supported edge along the y-axis is considered, **K** and **V** matrices need to be changed according to Eqs. (11.25) and (11.26).

$$\mathbf{K} = \begin{bmatrix} \mathbf{A} & \mathbf{B} & \mathbf{C} & \mathbf{0} & \cdots \\ \mathbf{B} & \mathbf{A} & \mathbf{B} & \mathbf{C} & \mathbf{0} \\ \mathbf{C} & \mathbf{B} & \mathbf{A} & \mathbf{B} & \mathbf{C} \\ \cdots & \cdots & \cdots & \cdots & \cdots \end{bmatrix} \text{ with } \mathbf{A} = \begin{bmatrix} 1 & 0 & 0 & 0 & \cdots \\ 1 & -2 & 1 & 0 & 0 \\ 1 & -8 & 20 & -8 & 1 \\ \cdots & \cdots & \cdots & \cdots & \cdots \end{bmatrix}$$

$$\mathbf{B} = \begin{bmatrix} 0 & 0 & 0 & 0 & \cdots \\ 0 & 0 & 0 & 0 & 0 \\ 0 & 2 & -8 & 2 & 0 \\ \cdots & \cdots & \cdots & \cdots & \cdots \end{bmatrix} \text{ and } \mathbf{C} = \begin{bmatrix} 0 & 0 & 0 & 0 & \cdots \\ 0 & 0 & 0 & 0 & 0 \\ 0 & 0 & 1 & 0 & 0 \\ \cdots & \cdots & \cdots & \cdots & \cdots \end{bmatrix}$$

$$(11.25)$$

$$\mathbf{V} = \begin{bmatrix} \mathbf{V}_1 \\ \cdots \\ \mathbf{V}_i \\ \cdots \\ \mathbf{V}_n \end{bmatrix} \text{ with } \mathbf{V}_i = \begin{bmatrix} 0 \\ 0 \\ \dfrac{q}{D} \\ \dfrac{q}{D} \\ \cdots \end{bmatrix} \qquad (11.26)$$

11.3.2.2 Clamped (Fixed) Edges

A clamped edge in a plate satisfies the conditions that both the deflection and its slope are zero at the clamped edge. For a clamped edge at a fixed x, all the grid points on this edge (at $x = x_{\text{clamped}}$) need to satisfy Eqs. (11.27) and (11.28).

$$w\big|_{x_{\text{clamped}}} = 0 \qquad (11.27)$$

$$\frac{\partial w}{\partial x}\bigg|_{x_{\text{clamped}}} = 0 \Rightarrow w(x_1) - w(x_0) = 0 \qquad (11.28)$$

To include these conditions, **K** and **V** matrices need to be changed, as shown in Eqs. (11.29) and (11.30).

$$\mathbf{K} = \begin{bmatrix} \mathbf{A}_{\text{lim1}} & \mathbf{B}_{\text{lim1}} & \mathbf{C}_{\text{lim1}} & \mathbf{0} & \cdots \\ \mathbf{B}'_{\text{lim2}} & \mathbf{A}_{\text{lim2}} & \mathbf{B}_{\text{lim2}} & \mathbf{C}_{\text{lim2}} & \mathbf{0} \\ \mathbf{C} & \mathbf{B} & \mathbf{A} & \mathbf{B} & \mathbf{C} \\ \cdots & \cdots & \cdots & \cdots & \cdots \end{bmatrix} \qquad (11.29)$$

with $\mathbf{A}_{\text{lim1}} = \mathbf{A}_{\text{lim2}} = \mathbf{I}_m$; $\mathbf{B}'_{\text{lim2}} = -\mathbf{I}_m$ and $\mathbf{C}_{\text{lim1}} = \mathbf{C}_{\text{lim2}} = \mathbf{B}_{\text{lim1}} = \mathbf{B}_{\text{lim2}} = \mathbf{0}$

$$\mathbf{V} = \begin{bmatrix} \mathbf{V}_1 \\ \ldots \\ \mathbf{V}_i \\ \ldots \\ \mathbf{V}_n \end{bmatrix} \text{ with } \mathbf{V}_i = \begin{bmatrix} \dfrac{q}{D} \\ \dfrac{\dot{q}}{D} \\ \dfrac{\dot{q}}{D} \end{bmatrix} \text{ and } \mathbf{V}_{1,2} = \begin{bmatrix} 0 \\ \ldots \\ \ldots \\ \ldots \\ 0 \end{bmatrix} \qquad (11.30)$$

If the clamped edge aligned along the y-axis is considered, Eqs. (11.27) and (11.28) need x to be replaced by y. This change modifies $\mathbf{K}$ and $\mathbf{V}$ matrices according to Eqs. (11.31) and (11.32).

$$\mathbf{K} = \begin{bmatrix} \mathbf{A} & \mathbf{B} & \mathbf{C} & \mathbf{0} & \ldots \\ \mathbf{B} & \mathbf{A} & \mathbf{B} & \mathbf{C} & \mathbf{0} \\ \mathbf{C} & \mathbf{B} & \mathbf{A} & \mathbf{B} & \mathbf{C} \\ \ldots & \ldots & \ldots & \ldots & \ldots \end{bmatrix} \text{ with } \mathbf{A} = \begin{bmatrix} 1 & 0 & 0 & 0 & \ldots \\ -1 & 1 & 0 & 0 & 0 \\ 1 & -8 & 20 & -8 & 1 \\ \ldots & \ldots & \ldots & \ldots & \ldots \end{bmatrix}$$

$$\mathbf{B} = \begin{bmatrix} 0 & 0 & 0 & 0 & \ldots \\ 0 & 0 & 0 & 0 & 0 \\ 0 & 2 & -8 & 2 & 0 \\ \ldots & \ldots & \ldots & \ldots & \ldots \end{bmatrix} \text{ and } \mathbf{C} = \begin{bmatrix} 0 & 0 & 0 & 0 & \ldots \\ 0 & 0 & 0 & 0 & 0 \\ 0 & 0 & 1 & 0 & 0 \\ \ldots & \ldots & \ldots & \ldots & \ldots \end{bmatrix}$$

$$(11.31)$$

$$\mathbf{V} = \begin{bmatrix} \mathbf{V}_1 \\ \ldots \\ \mathbf{V}_i \\ \ldots \\ \mathbf{V}_n \end{bmatrix} \text{ with } \mathbf{V}_i = \begin{bmatrix} 0 \\ 0 \\ \dfrac{q}{D} \\ \dfrac{q}{D} \\ \ldots \end{bmatrix} \qquad (11.32)$$

11.4 PLATE DEFLECTION PROBLEMS: NUMERICAL SOLUTIONS

A number of plate deflections problems have been solved using the finite difference method implemented in a spreadsheet program and the results are presented below [4].

11.4.1 Clamped Plate Under a Uniform Load

In the first case (Fig. 11.4), the problem of a uniformly loaded plate with its all edges clamped was considered. The plate was assumed to be a square plate made of aluminum with $L_1 = L_2 = 0.5$ m, $t = 0.01$ m, The material properties of aluminum were Young's modulus, $E = 70$ GPa, and Poisson's ratio $v = 0.3$, and the uniformly distributed load was $q = 0.5$ MPa. Several numerical models were solved by varying the number of grid points for the

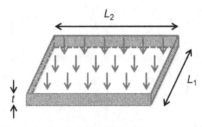

FIGURE 11.4 Schematic of a plate rigidly clamped subjected to a uniform load.

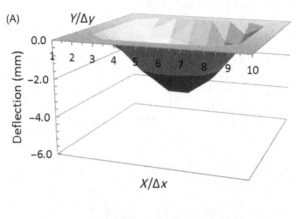

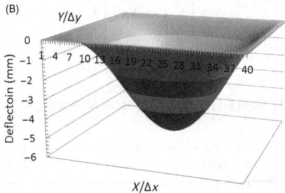

FIGURE 11.5 Deformed shape of the clamped plate with 10 grid points (A) and 40 grid points (B) along each side.

discretization, with $n = m = 5$, 10, 15, 25, 40. The results for the deformed configuration (deflected shape) of the plate for 10 and 40 points are shown in Fig. 11.5, which shows the deflection distribution over the domain of the plate.

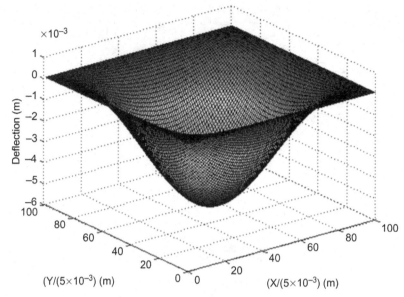

FIGURE 11.6 Deformed shape of the clamped plate with a 100 points discretization along each side.

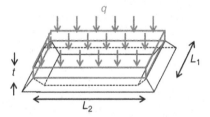

FIGURE 11.7 Schematic of a simply supported plate subjected to a uniform load.

These results are further discussed in Section 11.5. It can be seen that these results, however, do not agree well with the exact (analytical) solution. Next, to improve the accuracy of the numerical model, the number of grid points for the discretization was increased to a 75×75 and a 100×100 grid. The deformation configuration of the plate for the 100×100 grid is shown in Fig. 11.6. The accuracy of prediction has improved significantly with the fine resolution (i.e., increased grid points).

11.4.2 Simply Supported Plate Under a Uniform Load

Next a plate with simply supported conditions at the edges was considered (Fig. 11.7). The dimension and properties of the plate were the same as those

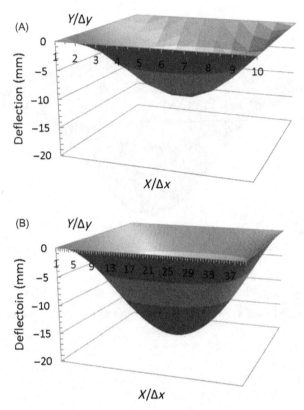

FIGURE 11.8 Deformed shape of the simply supported plate with 10 grid points (A) and 40 grid points (B) along each side.

of the clamped case, i.e., $L = L_1 = L_2 = 0.5$ m, $t = 0.01$ m, $E = 70$ GPa, $v = 0.3$, and the load intensity also was the same with $q = 0.5$ MPa.

Several models were analyzed with various discretized grid points. For all cases, the same number of grid points was used in the x and y directions ($n = m$) with the values being 10, 15, 25, and 40. Fig. 11.8 shows results for deflection of the plate for 10 and a 40 grid points.

To obtain improved accuracy, the same problem was solved with 100 grid points along each direction. Fig. 11.9 shows the deformed shape which is relatively smoother than the earlier low resolution results, and the results match closer to the analytical solution.

11.4.3 Simply Supported Plate Under a Concentrated Load

Next the same problem was solved with the loading changed to a concentrated load (Fig. 11.10). The geometry and material properties of the plate were the same as before, i.e., with those of the plates with uniformly

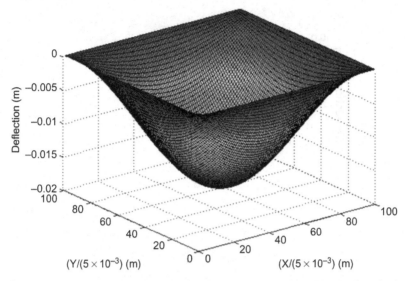

FIGURE 11.9   Deformed shape of the simply supported plate with a 100-point discretization along each side.

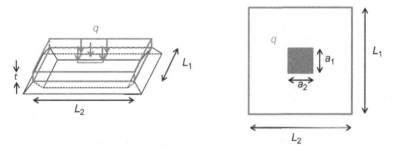

FIGURE 11.10   Schematic of a simply supported plate subjected to a concentrated (point) load.

distributed loading conditions. The load was concentrated at the center of the plate. In practice for implementation in the numerical method, the load was applied over a small area as shown in Fig. 11.10, with $a_1 = a_2 = 0.1$ m. This localized distributed load had a value of 0.5 MPa, making the equivalent concentrated load to be 5000 N.

The numerical solution method needed some adjustment in the right-hand side of the vector to take into account of the approximately concentrated load. The matrix **V** was to be modified setting $q = 0$ on all grid points outside the loaded region.

Several discretized models were solved with the number of grid points of 100 (10×10), 225 (15×15), 625 (25×25), and 1600 (40×40) points. Fig. 11.11 shows the results for the deflected plate shapes for 100 (10×10) and 1600 (40×40) discretized grids.

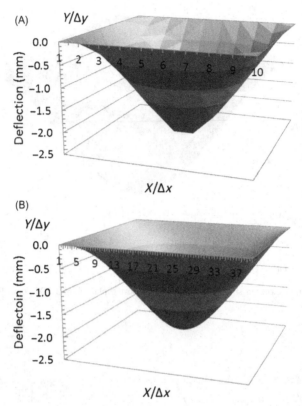

FIGURE 11.11 Deformed shape of the simply supported plate under concentrated load with 10 grid points (A) and 40 grid points (B) along each side.

11.4.4 Simply Supported Plate Under a Uniform Load With a Linear Stiffness Variation

The numerical modeling work was then extended to a "functionally graded material." This is a more complex problem where the material property varies continuously along a specific direction. For such problems, the analytical solution is quite complex. In this case, the Young's modulus of the plate was varied linearly along both edges (both in the x and y directions). The variation of the Young's modulus (E) is given by Eq. (11.33).

$$E(x, y) = \frac{x}{L}\left(E_{x\_max} - E_{x\_min}\right) + \frac{y}{L}\left(E_{y\_max} - E_{y\_min}\right) + E_{x\_min} + E_{y\_min}$$

(11.33)

where $E(x, y)$ is the Young's modulus at a given location (x, y), E_{x_min} and E_{x_max} are the minimum and maximum values of E from one edge to the other edge along the x-axis varying linearly, and E_{y_min} and E_{y_max} are the

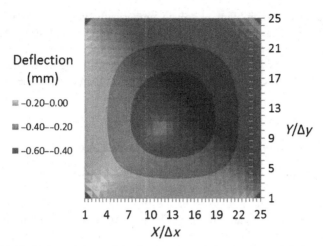

FIGURE 11.12 Deformed shape of the simply supported functionally graded plate with linear variation of Young's modulus under uniform load with 25 grid points along each side.

corresponding values along the y-axis. For the numerical solution, the matrix **V** needs to be changed in (Eqs. 11.18 and 11.24) to take into account of the variation of material property, and the parameter D needs to be modified to incorporate the variation of $E(x, y)$ as given in Eq. (11.34).

$$D_{i,j} = \frac{E_{i,j}t^3}{12(1 - v^2)} \tag{11.34}$$

It is important to understand the increment of matrix **K** to follow the change of matrix **V**. The same plate problem, i.e., the plate with simply supported edges as shown in Section 11.4.2 is studied here, however, with a linear variation of Young's modulus, given by Eq. (11.35), and subjected to the same uniform load of $q = 0.5$ MPa. Fig. 11.12 shows the deflection of the plate for a 625 grid point (25×25 points) discretization.

$$E(x, y) = 5000\left(1 + \frac{y}{L}\right) \text{ MPa} \tag{11.35}$$

It is to be noted that the deformation pattern of the plate for functionally graded Young's modulus is different from a homogeneous plate with constant material property. This also demonstrates the advantage of numerical methods for finding solution of a relatively complex problem where a simple analytical solution does not exist or is difficult to determine.

11.5   CONVERGENCE AND ACCURACY

In this section, the convergence and accuracy of the numerical methods in relation to plate bending problems are investigated. For all of the plate

250 Demystifying Numerical Models

configurations, the problem is solved numerically with progressively finer resolution grids, i.e., increasing grid points or smaller grid sizes. The solutions are successively compared so that the results obtained for a specific grid is compared with those obtained from the next fine grid. If an analytical solution exists, then these results are also compared with the analytical solution.

11.5.1 Clamped Plate Under a Uniform Load

The primary parameter of the deflected shape considered for comparison of the numerical solutions is the vertical deflection at the center point of the plate ($w_{\max}$). This is used to calculate the difference or error between the numerical and analytical solutions. The analytical solution of deflection at any point (x,y) of a clamped square plate under a uniform load is given by Eq. (11.36). The maximum center point deflection, which is of interest, is given by Eq. (11.37). These equations are valid for a square plate with edges clamped and $v = 0.3$ [3].

$$w(x,y) = \frac{16q}{\pi^6 D} \sum_m^\infty \sum_n^\infty \frac{\sin\left(\frac{m\pi x}{L}\right)\sin\left(\frac{n\pi y}{L}\right)}{mn\left(\frac{(m^2+n^2)}{L^2}\right)^2} \tag{11.36}$$

with m and n being odd integers.

$$w_{\max} = 0.00126\frac{qL^4}{D} \tag{11.37}$$

For the geometry and property of the plate considered in Section 11.4.1, the analytical solution provides the exact value, $w_{\max,ana} = -6.14$ mm. The problem was solved numerically using the finite difference method for various grid resolutions, as given by the number of points along each side of the plate. Table 11.1 summarizes the results including the maximum center point deflections, the errors compared to the analytical solution, and the solution times for various grid resolutions. The error is given by the difference between the analytical and numerical solutions expressed as a percentage of the analytical solution. The solution time was the computer (CPU) time required to run the problem using the spreadsheet program (Microsoft Excel). The numerical solutions show that for coarse (low resolution) grids with 5, 10, and 15 points per edge, the solution accuracy is poor, with the corresponding errors being in the range of 27%–85%. With an increase in grid resolution, i.e., as more points were used to discretize the plate surface, the solution approaches close to the analytical solution, as demonstrated by

TABLE 11.1 Effect of Discretization (Number of Points Per Edge) on the Maximum Deflection, Error and Solution Time for the Clamped Plate Under Uniform Load

Number of Points Per Edge	w_{max} (mm)	Error (%)	Solution Time (s)
5	− 0.952	84.5	0.10
10	− 3.33	45.8	0.11
15	− 4.44	27.8	0.12
25	− 5.15	16.2	0.13
40	− 5.53	10.0	0.37
60	− 5.70	6.5	3.30
75	− 5.80	5.0	9.75
85	− 5.90	4.4	20.14
90	− 5.90	4.1	29.30
95	− 5.90	3.8	38.00
100	− 5.90	3.7	49.86

the considerable reduction in error. The decrease in the error with an increase in the number of grid points (per side of the plate) is shown in Fig. 11.13. When more than 75 points were used, the error was reduced below 5%, indicating a very good accuracy of prediction of the maximum deflection. At grid point resolution of 100 points, the maximum deflection was −5.90 mm, which is very close to the exact value (an error of only 3.7%). However the solution time increases with increasing grid size as shown in Fig. 11.14. From these results, it is often recommended to choose a balance between accuracy and solution time.

It is possible to find the trend lines for the variation of the error and time for solution (see Figs. 11.13 and 11.14). Using these equations, it is possible to estimate the solution time and the error for a given discretization.

11.5.2 Simply Supported Plate Under a Uniform Load

The same problem, as in Section 11.4.2, was numerically solved using different grid resolutions, ranging from 10 to 100, along each side. As in the previous problem, the exact analytical solution of the maximum deflection of the center point for this case is given by Eq. (11.38), and it is used for

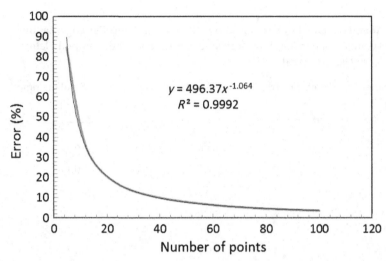

FIGURE 11.13 Effect of discretization (number of points) on the error.

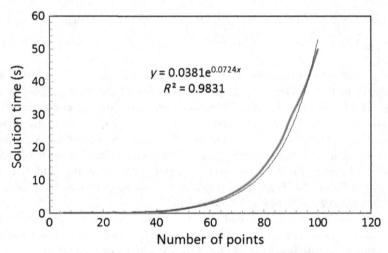

FIGURE 11.14 Effect of discretization (number of points) on the solution time.

comparison against the numerical solutions. This solution is for a square plate with simply supported edge condition and with $v = 0.3$ [3]. With the same geometry, material, and loading, the analytical solution provides a maximum vertical deflection of $w_{max} = -19.8$ mm.

$$w_{max} = 0.00406 \frac{qL^4}{D} \qquad (11.38)$$

TABLE 11.2 Effect of Discretization (Number of Points Per Edge) on the Maximum Deflection, Error and Solution Time for the Simply Supported Plate Under Uniform Load

Number of Points Per Edge	w_{max} (mm)	Error (%)	Solution Time (s)
10	−11.6	41.5	0.10
15	−14.8	25.2	0.11
25	−16.9	14.5	0.12
40	−18.0	9.1	0.38
60	−18.6	5.8	3.01
75	−18.9	4.6	10.84
85	−19.0	4.0	18.72
90	−19.0	3.8	29.50
95	−19.1	3.6	40.60
100	−19.1	3.4	50.38

The results from the numerical method, including the maximum deflections, the errors, and the solution times, are summarized in Table 11.2. Similar to the clamped case, the error (difference from exact solution) is large at low resolutions. For example, the error is >25% when the number of grid points along each side is about 15. The error decreases rapidly with the increase in the number of grid points. The error becomes <10% with the number of points being 40 along each side. The accuracy of the numerical solutions is very high when the number of points exceeds 75. A fine grid of 100 grid points along each side produces a maximum deflection of −19.1 mm, which is very close to the analytical value of −19.8 mm, with the error being 3.4%.

From these results, the following empirical equations can be derived by fitting of the data. Eq. (11.39) provides the relationship between the error (*Err*) and the number of grid points (*n*), and the relationship between the solution time (t_{sol}) and the number of grid points (*n*) is given by Eq. (11.40). These equations can assist in selecting a suitable grid resolution to ensure a balance between accuracy and computer time.

$$Err = 474.2n^{-1.074} \qquad (11.39)$$

$$t_{\text{sol}} = 0.0305e^{0.0753n} \qquad (11.40)$$

11.5.3 Simply Supported Plate Under a Concentrated Load

The same problem of the simply supported plate under a concentrated load, as in Section 11.4.3, was numerically solved using different grid points. The corresponding problem has an analytical expression using Navier solution given by Eq. (11.41) [3]. It is to note that the analytical solution assumes a perfectly concentrated or a point load, whereas in the numerical models, the load was applied over a small area for practical purpose.

$$w_{\text{max}} = \frac{FL^2}{2\pi^3 D} \sum_{m=1}^{\infty} \frac{1}{m^3} \left(\tanh(\alpha_m) - \frac{\alpha_m}{\cosh^2(\alpha_m)} \right)$$

$$\alpha_m = m\frac{\pi}{2} \text{ with } m \text{ is a odd integer} \qquad (11.41)$$

This series given by Eq. (11.41) can be approximated using only the first terms to render approximate Eq. (11.42) for the center point (maximum) deflection of the plate.

$$w_{\text{max}} = 0.01160 \frac{FL^2}{D} \qquad (11.42)$$

Similar to the previous two cases of plate deflections, the accuracy of the numerical FD models was very good as compared with this analytical solution, with the error being <4% for a number of grid points along each side more than 20.

11.6 SUMMARY

This chapter deals with an important category of structural component used in shell type structures, i.e., the plate structures. Plates and shells are commonly used in many structural applications, e.g., aerospace, automotive, marine, and machinery. This chapter demonstrates that the FDM is a useful tool for accurate analysis of plate structures under different loads.

It can be implemented in a spreadsheet program (e.g., Microsoft Excel) to solve for plate problems numerically by applying suitable boundary conditions. The solutions were found to converge to the respective analytical solutions with a reasonable grid point resolution, ascertaining the accuracy of the method. Most importantly, for complex plate structures, particularly those integrated with other sub-structures resulting in complex geometry and boundary conditions, an analytical solution often does not exist or difficult to obtain. Such problems can be effectively solved by simple use of the FDM implemented in a spreadsheet program.

REFERENCES

[1] Timoshenko S, Goodier JN. Theory of elasticity. New York: McGraw-Hill Book Company; 1951.

[2] Megson THG. Chapter 7 – Bending of thin plates. *Introduction to aircraft structural analysis.* Boston: Butterworth-Heinemann; 2010. p. 219–52.

[3] Timoshenko S, Woinowsky-Krieger S. Theory of plates and shells. New York: McGraw-Hill Book Company; 1989.

[4] Application of Numerical Methods for the Solution of Solid Mechanics Problems. Pantaloni, D., and Das, R., 2017. RMIT University, Australia, Report No. 2017/2A-SIM.

Index

Note: Page numbers followed by "*f*" and "*t*" refer to figures and tables, respectively.

Printed in the United States
By Bookmasters